Robust Nonlinear Regression

Robust Nonlinear Regression:
with Applications using R

Hossein Riazoshams
Lamerd Islamic Azad University, Iran
Stockholm University, Sweden
University of Putra, Malaysia

Habshah Midi
University of Putra, Malaysia

Gebrenegus Ghilagaber
Stockholm University, Sweden

Registered Offices
John Wiley & Sons, Inc., 111 River Street, Hoboken, NJ 07030, USA
John Wiley & Sons Ltd, The Atrium, Southern Gate, Chichester, West Sussex, PO19 8SQ, UK

Editorial Office
9600 Garsington Road, Oxford, OX4 2DQ, UK

For details of our global editorial offices, customer services, and more information about Wiley products visit us at www.wiley.com.

Wiley also publishes its books in a variety of electronic formats and by print-on-demand. Some content that appears in standard print versions of this book may not be available in other formats.

Library of Congress Cataloging-in-Publication Data

Names: Riazoshams, Hossein, 1971– author. | Midi, Habshah, author. |
    Ghilagaber, Gebrenegus, author.
Title: Robust nonlinear regression: with applications using R / Hossein
    Riazoshams, Habshah Midi, Gebrenegus Ghilagaber.
Description: Hoboken, NJ : John Wiley & Sons, 2018. | Includes
    bibliographical references and index. |
Identifiers: LCCN 2017057347 (print) | LCCN 2018005931 (ebook) | ISBN
    9781119010456 (pdf) | ISBN 9781119010449 (epub) | ISBN 9781118738061
    (cloth)
Subjects: LCSH: Regression analysis. | Nonlinear theories. | R (Computer
    program language)
Classification: LCC QA278.2 (ebook) | LCC QA278.2 .R48 2018 (print) | DDC
    519.5/36–dc23
LC record available at https://lccn.loc.gov/2017057347

Cover Design: Wiley
Cover Image: © Wavebreakmedia Ltd/Getty Images; © Courtesy of Hossein Riazoshams

Set in 10/12pt WarnockPro by SPi Global, Chennai, India
Printed in Singapore by C.O.S. Printers Pte Ltd

10 9 8 7 6 5 4 3 2 1

To my wife Benchamat Hanchana, from Hossein

Contents

Preface

This book is the result of the first author's research, between 2004 and 2016, in the robust nonlinear regression area, when he was affiliated with the institutions listed. The lack of computer programs together with mathematical development in this area encouraged us to write this book and provide an R-package called `nlr` for which a guide is provided in this book. The book concentrates more on applications and thus practical examples are presented.

Robust statistics describes the methods used when the classical assumptions of statistics do not hold. It is mostly applied when a data set includes outliers that lead to violation of the classical assumptions.

The book is divided into two parts. In Part 1, the mathematical theories of robust nonlinear regression are discussed and parameter estimation for heteroscedastic error variances, autocorrelated errors, and several methods for outlier detection are presented. Part 2 presents numerical methods and R-tools for nonlinear regression using robust methods.

In Chapter 1, the basic theories of robust statistics are discussed. Robust approaches to linear regression and outlier detection are presented. These mathematical concepts of robust statistics and linear regression are then extended to nonlinear regression in the rest of the book. Since the book is about nonlinear regression, the proofs of theorems related to robust linear regression are omitted.

Chapter 2 presents the concepts of nonlinear regression and discusses the theory behind several methods of parameter estimation in this area. The robust forms of these methods are outlined in Chapter 3. Chapter 2 presents the generalized least square estimate, which will be used for non-classical situations.

Chapter 3 discusses the concepts of robust statistics, such as robustness and breakdown points, in the context of nonlinear regression. It also presents several robust parameter estimation techniques.

Chapter 4 develops the robust methods for a null condition when the error variances are not homogeneous. Different kinds of outlier are defined and their effects are discussed. Parameter estimation for nonlinear function models and variance function models are presented.

Another null condition, when the errors are autocorrelated, is discussed in Chapter 5. Robust and classical methods for estimating the nonlinear function model and the autocorrelation structure of the error are presented. The effect of different kinds of outlier are explained, and appropriate methods for identifying the correlation structure of errors in the presence of outliers are studied.

Chapter 6 explains the methods for identifying atypical points. The outlier detection methods that are developed in this chapter are based mainly on statistical measures that use robust estimators of the parameters of the nonlinear function model.

In Chapter 7, optimization methods are discussed. These techniques are then modified to solve the minimization problems found in robust nonlinear regressions. They will then used to solve the mathematical problems discussed in Part 1 of the book and their implementation in a new R package called nlr is then covered in Chapter 8.

Chapter 8 is a guide to the R package implemented for this book. It covers object definition for a nonlinear function model, parameter estimation, and outlier detection for several model assumption situations discussed in the Part 1. This chapter shows how to fit nonlinear models to real-life and simulated data.

In Chapter 9, another R packages for robust nonlinear regression are presented and compared to nlr. Appendix A presents and describes the databases embedded in nlr, and the nonlinear models and functions available.

At the time of writing, the nlr package is complete, and is available at The Comprehensive R Archive Network (CRAN-project) at https://cran.r-project.org/package=nlr.

Because of the large number of figures and programs involved, there are many examples that could not be included in the book. Materials, programs, further examples, and a forum to share and discuss program bugs are all provided at the author's website at http://www.riazoshams.com/nlr and at the book's page on the Wiley website.

Response Manager, Shabdiz Music School of Iran,
Full time faculty member of Islamic Azad University of Lamerd, Iran,
Department of Statistics, Stockholm University, Sweden,
Institute for Mathematical, Research University of Putra, Malaysia
November 2017

Hossein Riazoshams

Acknowledgements

I would like to thank the people and organizations who have helped me in all stages of they research that has culminated in this book. Firstly I would like to express my appreciation to Mohsen Ghodousi Zadeh and Hamid Koohbor for helping me in collecting data for the first time in 2005. This led me to a program of research in nonlinear modeling.

I would like to recognize the Department of Statistics at Stockholm University, Sweden, for financial support while writing most of this book during my stay as a post-doctoral researcher in 2012–2014.

A special note of appreciation is also due to the Islamic Azad University of Abadeh and Lamerd for financial support in connection with collecting some materials for this book.

I would like note my appreciation for the Institute for Mathematical Research of University Putra Malaysia for financial support during my PhD in 2007–2010 and afterwards.

I owe my gratitude to the John Wiley editing team, specially Shyamala and others for their great editing process during the preparation of the book.

Last but by no means least, I would like to thank my wife, Benchamat Hanchan, for her great patience with the financial and physical adversity that we experienced during this research.

November 2017   *Hossein Riazoshams*

About the Companion Website

Don't forget to visit the companion website for this book:

www.wiley.com/go/riazoshams/robustnonlinearregression

There you will find valuable material designed to enhance your learning, including:

- Figures
- Examples

Scan this QR code to visit the companion website

Part One

Theories

1

Robust Statistics and its Application in Linear Regression

This is an introductory chapter giving the mathematical background to the robust statistics that are used in the rest of the book. Robust linear regression methods are then generalized to nonlinear regression in the rest of the book.

The robust approach to linear regression is described in this chapter. It is the main motivation for extending statistical inference approaches used in linear regression to nonlinear regression. This is done by considering the gradient of a nonlinear model as the design matrix in a linear regression. Outlier detection methods used in linear regression are also extended to use in nonlinear regression.

In this chapter the consistency and asymptotic distributions of robust estimators and robust linear regression are presented. The validity of the results requires certain regularity conditions, which are presented here. Proofs of the theorems are very technical and since this book is about nonlinear regression, they have been omitted.

1.1 Robust Aspects of Data

Robust statistics were developed to interpret data for which classical assumptions, such as randomness, independence, distribution models, prior assumptions about parameters and other prior hypotheses do not apply. Robust statistics can be used in a wide range of problems.

The classical approach in statistics assumes that data are collected from a distribution function; that is, the observed values $(x_1, x_2, ..., x_n)$ follow the simultaneous distribution function $F_n(x_1, ..., x_n)$. If the observations are identically independently distributed (i.i.d.) with distribution F, we write $x_i \overset{iid}{\sim} F(x_i), i = 1, ..., n$ (the tilde sign $\sim$ designates a distribution). In real-life data, these explicit or other implicit assumptions might not be true. Outlier data effects are examples of situations that require robust statistics to be used for such null conditions.

Robust Nonlinear Regression: with Applications using R, First Edition.
Hossein Riazoshams, Habshah Midi, and Gebrenegus Ghilagaber.
© 2019 John Wiley & Sons Ltd. Published 2019 by John Wiley & Sons Ltd.
Companion website: www.wiley.com/go/riazoshams/robustnonlinearregression

1.2 Robust Statistics and the Mechanism for Producing Outliers

Robust statistics were developed to analyse data drawn from wide range of distributions and particularly data that do not follow a normal distribution, for example when a normal distribution is mixed with another known statistical distribution:[1]

$$F = (1 - \varepsilon)N(:, :) + \varepsilon D \tag{1.1}$$

where ε is a small value representing the proportion of outliers, $N(:, :)$ is the normal cumulative distribution function (CDF) with appropriate mean and variance, and D belongs to a suitable class of CDFs. A normal distribution (D) with a large variance can produce a wide distribution, such as:

$$F = (1 - \varepsilon)N(0, \sigma^2) + \varepsilon N(0, \sigma_0^2)$$

for a large value of σ_0^2 (see Figure 1.1a). A mixture of two normal distributions with a large difference in their means can be generated by:

$$F = (1 - \varepsilon)N(0, \sigma^2) + \varepsilon N(\mu_0, \sigma_0^2)$$

where the variance value σ_0^2 is much smaller than σ^2, and the mean μ_0 is the mean of the shifted distribution (see Figure 1.1b). The models in this book will

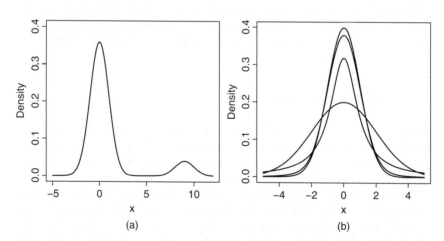

Figure 1.1 Contaminated normal densities: (a) mixture of two normal distributions with different means; (b) mixture of two normal distributions with different variances. *Source*: Maronna et al. (2006). Reproduced with permission of John Wiley and Sons.

---

1 Some of the mathematical notation and definitions used in this chapter follow that in Maronna, et al. (2006).

be used to interpret data sets with outliers. Figure 1.1a shows the CDF of a mixture of two normal distributions with different means:

$$F = 0.9 \times N(0, 1) + 0.1 \times N(9, 1)$$

and Figure 1.1b shows the CDF of a mixture of two normal distributions with different variances:

$$F = 0.9 \times N(0, 1) + 0.1 \times N(0, 2)$$

1.3 Location and Scale Parameters

In this section we discuss the location and scale models for random sample data. In later chapters these concepts will be extended to nonlinear regression. The location model is a nonlinear regression model and the scale parameter describes the nonconstant variance case, which is common in nonlinear regression.

1.3.1 Location Parameter

Nonlinear regression, and linear regression in particular, can be represented by a location model, a scale model or simultaneously by a location model and a scale model (Maronna et al. 2006). Not only regression but also many other random models can be systematically studied using this probabilistic interpretation. We assume that an observation $x_i, i = 1, \ldots, n$ depends on the unknown true value μ and that a random process acts additively as

$$x_i = \mu + \varepsilon_i, \ i = 1, \ldots, n \tag{1.2}$$

where the errors ε_i are random variables. This is called the *location model* and was defined by Huber (1964). If the errors ε_i are independent with common distribution F_0 then the x_i outcomes are independent, with common distribution function

$$F(x) = F_0(x - \mu)$$

and density function $f_0(x - \mu) = F_0'$. An estimate $\hat{\mu}$ is a function of the observations $\hat{\mu} = \hat{\mu}(x_1, \ldots, x_n)$. We are looking for estimates that, with high probability, satisfy $\hat{\mu} \simeq \mu$. The *maximum likelihood estimate* (MLE) of μ is a function of observations that maximize the likelihood function (joint density):

$$L(x_1, \ldots, x_n; \mu) = \prod_{i=1}^{n} f_0(x_i - \mu) \tag{1.3}$$

The estimate of the location can be obtained from:

$$\hat{\mu}_{MLE} = \hat{\mu}(x_1, \ldots, x_n) = \arg \max_{\mu} L(x_1, \ldots, x_n; \mu)$$

Since f_0 is positive and the logarithm function is an increasing function, the MLE of a location can be calculated using a simple maximization logarithm statement:

$$\hat{\mu}_{MLE} = \arg \max_{\mu} \ell(\mu)$$

$$= \arg \max_{\mu} \sum_{i=1}^{n} \log(f_0(x_i - \mu)) \tag{1.4}$$

If the distribution F_0 is known then the MLE will have desirable mathematical and *optimality* properties, in the sense that among unbiased estimators it has the lowest variance and an approximately normal distribution. In the presence of outliers, since the distribution F_0 and, in particular, the mixture distribution (1.1) are unknown or only approximately known, statistically *optimal* properties might not be achieved. In this situation, some optimal estimates can still be found, however. Maronna et al. (2006, p. 22) state that to achieve optimality, the goal is to find estimates that are:

- *nearly optimal* when F_0 is normal
- *nearly optimal* when F_0 is approximately normal.

To this end, since MLEs have good properties such as sufficiency, known distribution and minimal bias within an unbiased estimator but are sensitive to the distribution assumptions, an MLE-type estimate of (1.4) can be defined. This is called an M-estimate. As well as the M-estimate for location, a more general definition can be developed. Let:

$$\rho(t) = -\log f_0(t) \tag{1.5}$$

The negative logarithm of (1.3) can then be written as $\sum_{i=1}^{n} \rho(x_i - \mu)$.

A more sophisticated form of M-estimate can be defined by generalizing to give an estimator for a multidimensional unknown parameter μ of an arbitrary modeling of a given random sample $(x_1, \ldots, x_n)$.

Definition 1.1 If a random sample $(x_1, \ldots, x_n)$ is given, and $\mu \in \Re^p$ is an unknown p-dimensional parameter of a statistical model describing the behavior of the data, any estimator of μ is a function of a random sample $\hat{\mu} = \mu(x_1, \ldots, x_n)$. The M-estimate of μ can be defined in two different ways: by a minimization problem of the form (estimating equation and functional form are represented together):

$$\sum_{i=1}^{n} \rho(x_i, \hat{\mu}) = \int \rho(x, \hat{\mu}) dF_n = \min \tag{1.6}$$

or as the solution of the equation with the functional form

$$\sum_{i=1}^{n} \psi(x_i; \hat{\mu}) = \int \psi(x, t) dF_n(x) = 0 \tag{1.7}$$

where the functional form means $E(\psi(x, t)) = 0$, F_n is an empirical CDF, and ρ (the robust loss function) and ψ are arbitrary functions. If ρ is partially differentiable, we can define the psi function as $\psi_{p \times 1}(x_i; \boldsymbol{\mu}) = (\partial/\partial \boldsymbol{\mu})\rho(x_i; \boldsymbol{\mu})$, which is specifically proportional to the derivative ($\psi \propto \rho'$), and the results of Equations 1.6 and 1.7 are equal. In this section we are interested in the M-estimate of the location for which $\psi(x, t) = \psi(x - t)$.

The M-estimate was first introduced for the location parameter by Huber (1964). Later, Huber (1972) developed the general form of the M-estimate, and the mathematical properties of the estimator (1973; 1981).

Definition 1.2 The M-estimate of location μ is defined as the answer to the minimization problem:

$$\hat{\mu} = \arg \min_{\mu} \sum_{i=1}^{n} \rho(x_i - \mu) \tag{1.8}$$

or the answer to the equation:

$$\sum_{i=1}^{n} \psi(x_i - \hat{\mu}_n) = 0 \tag{1.9}$$

If the function ρ is differentiable, with derivative $\psi(t) = d\rho/dt$, the M-estimate of the location (1.8) can be computed from the implicit equation (1.9).

If F_0 is a normal distribution, the ρ function, ignoring constants, is a quadratic function $\rho(x) = x^2$ and the parameter estimate is equivalent to the least squares estimate, given by:

$$\hat{\mu} = \arg \min_{\mu} \sum_{i=1}^{n} (x_i - \mu)^2$$

which has the average solution $\hat{\mu} = \bar{x} = (1/n) \sum_{i=1}^{n} x_i$.

If F_0 is a double exponential distribution with density $f_0(x) = (1/2)e^{-|x|}$, the rho function, apart from constants, is the absolute value function $\rho(x) = |x|$, and the parameter estimate is equivalent to the least median estimate given by:

$$\hat{\mu} = \arg \min_{\mu} \sum_{i=1}^{n} |x_i - \mu| \tag{1.10}$$

which has median solution $\hat{\mu} = median(x_i)$ (see Exercise 1). Apart from the mean and median, the distribution of the M-estimate is not known, but the convergence properties and distribution can be derived. The M-estimate is defined under two different formulations: the ψ approach from the estimating equation $\sum_{i=1}^{n} \psi(x_i, \mu) = 0$ or by minimization of $\sum_{i=1}^{n} \rho(x_i, \mu)$, where ρ is a primitive function of ψ with respect to μ. The consistency and asymptotic assumptions of the M-estimate depend on a variety of assumptions. The ψ approach does not have a unique root or an exact root, and a rule is required for selecting a root when multiple roots exist.

Theorem 1.3 Let $\lambda_F(t) = \int \psi(t, x) dF(x) = 0$. Assume that:

A 1.4

(i) $\lambda_{F_n(t)} = 0$ has unique root $\mu_0 = 0$
(ii) ψ is continuous an either bounded or monotone.

Then the equation $\lambda_{F_n(t)} = 0$ has a sequence of roots $\hat{\mu}_n$ that converge in probability $\hat{\mu}_n \xrightarrow{a.s.} \mu_0$.

In most cases, the equation $\lambda_{F_n(t)} = 0$ does not have an explicit answer and has to be estimated using numerical iteration methods. Starting from $\sqrt{n}$ consistent estimates $\hat{\mu}_n$, one step of the Newton–Raphson estimate is $\delta_n = \hat{\mu}_n - \frac{\psi_{F_n}(\hat{\mu}_n)}{\psi'_{F_n}(\hat{\mu}_n)}$, where $\psi'_{F_n}(t) = (\partial/\partial t) \sum \psi(x_i - t)$. The consistency and normality of δ_n are automatic, but there is no root of $\lambda_{F_n} = 0$ and furthermore the iteration does not change the first-order asymptotic properties.

Theorem 1.5 Suppose the following assumptions are satisfied:

A 1.6

(i) $\lambda_{F(t)} = 0$ has unique root t_0
(ii) $\psi(x, t)$ is monotone in t
(iii) $\lambda'_F(t_0)$ exists and $\neq 0$
(iv) $\int_{t_0} \psi^2(x, t) dF(x) < \infty$ in some neighborhood of t_0 and is continuous at t_0.

Then any sequence of the root of $\lambda_F(t) = 0$ satisfies

$$\sqrt{n}(\hat{\mu} - t_0) \xrightarrow{D} N\left(0, \frac{\int \psi^2(x, t_0) dF(x)}{(\int \psi'(x, t) dF(x))^2|_{t=t_0}}\right) \tag{1.11}$$

For a proof, see DasGupta (2008) and Serfling (2002).

Thus the location estimate μ from (1.9) or (1.8), under the conditions of Theorem 1.3, will converge in probability to the exact solution of $\sum \psi(x_i - \mu_0) = 0$ as $n \to \infty$. Under the conditions of Theorem 1.5 it will have a normal distribution:

$$\mu \sim N\left(\mu_0, \frac{v}{n}\right) \text{ with } v = \frac{E_F(\psi(x - \mu_0)^2)}{(E_F\psi'(x - \mu_0))^2} \tag{1.12}$$

In Equation (1.12), the parameter μ_0 is unknown and one cannot calculate the variance of the estimate. Instead, for inference purposes, we replace the expectations in the equation with the average, and parameter μ_0 by its estimate $\hat{\mu}_n$:

$$\hat{v} = \frac{\sum \psi(x_i - \hat{\mu}_n)^2}{\sum [\psi'(x_i - \hat{\mu}_n)]^2} \tag{1.13}$$

In Appendix A.3 several robust loss functions are presented.

1.3.2 Scale Parameters

In this section we discuss the scale model and parameters. The scale parameter estimate value is not only important in applications, but also plays a crucial role in computational iterations and the heteroscedasticity of variance cases. Consider the observations x_i that satisfy the multiplicative model known as the scale model:

$$x_i = \sigma \varepsilon_i \tag{1.14}$$

The values ε_i are i.i.d., with density f_0, and $\sigma > 0$ is an unknown parameter called the scale parameter. The distribution of x_i follows the scale family:

$$\frac{1}{\sigma} f_0 \left(\frac{x}{\sigma} \right).$$

Examples are the exponential family $f_0(x) = exp(-x)I(x > 0)$ and the normal scale family $N(0, \sigma^2)$.

Thus the MLE of σ is:

$$\hat{\sigma} = \arg\max_{\sigma} \frac{1}{\sigma^n} \prod_{i=1}^{n} f_0(x_i/\sigma)$$

Let $e_i = x_i/\sigma$. Taking logs and differentiating with respect to σ yields

$$\frac{1}{n} \sum_{i=1}^{n} -\frac{f_0'(e_i)}{f_0(e_i)} \frac{x_i}{\sigma} = 1$$

Let $\rho(t) = t\psi(t)$. If $\psi(t) = -f_0'(t)/f_0$, the estimating equation for σ can be written as

$$\frac{1}{n} \sum_{i=1}^{n} \rho \left(\frac{x_i}{\hat{\sigma}} \right) = 1 \tag{1.15}$$

Definition 1.7   Assume in the multiplicative model that $(x_1 \ldots, x_n)$ are n i.i.d. random samples of random variable X and $\varepsilon_1, \ldots, \varepsilon_n$ are n random samples of error random variable ε. X follows the multiplicative model $X = \sigma\varepsilon$.

For an appropriate ρ function and constant δ the M-estimate of scale σ is defined as (Huber 1964)

$$\int \rho\left(\frac{x}{\sigma(F)}\right) dF = E\left(\frac{x}{\sigma(F)}\right) = \delta \tag{1.16}$$

with sequence of estimation

$$\frac{1}{n}\sum_{i=1}^{n}\rho\left(\frac{x_i}{\sigma}\right) = \delta \tag{1.17}$$

Under regularity conditions, this converges to the functional form of (1.16).

1.3.3 Location and Dispersion Models

In an alternative approach, the location–dispersion model with two unknown parameters is defined as:

$$x_i = \mu + \sigma\varepsilon_i, \ i = 1, \dots, n \tag{1.18}$$

where ε has density f_0 and hence x_i has density

$$f(x) = \frac{1}{\sigma}f_0\left(\frac{x - \mu}{\sigma}\right)$$

In this case, σ is a scale parameter of $\sigma\varepsilon_i$, but a dispersion parameter of x_i. In practice, the parameter estimate for μ depends on σ, which might be known or unknown. The MLE for estimating (μ, σ) simultaneously is:

$$(\hat{\mu}, \hat{\sigma}) = \arg\max_{\mu,\sigma} \frac{1}{\sigma^n}\prod_{i=1}^{n}f_0\left(\frac{x_i - \mu}{\sigma}\right) \tag{1.19}$$

which, after taking logs and changing the sign, can be written as an optimization problem:

$$(\hat{\mu}, \hat{\sigma}) = \arg\min_{\mu,\sigma}\left\{\frac{1}{n}\sum_{i=1}^{n}\rho_0\left(\frac{x_i - \mu}{\sigma}\right) + \log(\sigma)\right\} \tag{1.20}$$

with $\rho_0 = -\log f_0$. Equations 1.19 and 1.20 are extensively used in this book to develop the underlying theory and also in reaching parameter estimates in problems of nonlinear and robust nonlinear regression. By calculating the derivative of (1.20) with respect to location and dispersion, the MLE of simultanous (μ, σ) can be defined, as in 1.9 and 1.17, by the simultaneous equations:

$$\begin{cases} \text{estimating } \mu \ : \ \sum_{i=1}^{n}\psi\left(\frac{x_i-\hat{\mu}}{\hat{\sigma}}\right) = 0 & \text{(1.21a)} \\[2mm] \text{estimating } \sigma \ : \ \frac{1}{n}\sum_{i=1}^{n}\rho_{\text{scale}}\left(\frac{x_i-\hat{\mu}}{\hat{\sigma}}\right) = \delta & \text{(1.21b)} \end{cases}$$

where $\psi(t) = -\rho_0'(t)$, $\rho_{\text{scale}}(t) = t\psi(t)$ and $\delta = 1$. The functional form can be written as:

$$\begin{cases} \text{estimating } \mu \; : \; E\psi\left(\frac{x-\mu}{\sigma}\right) = 0 & (1.22a) \\ \text{estimating } \sigma \; : \; E\rho_{\text{scale}}\left(\frac{x-\mu}{\sigma}\right) = \delta & (1.22b) \end{cases}$$

It can be proved that if F is symmetric then $\hat{\mu} \sim N(\mu, v/n)$, where:

$$v = \sigma^2 \frac{E[\psi(x-\mu)/\sigma]^2}{[E\psi(x-\mu)/\sigma]^2}$$

For computational purposes, the expectation can be estimated by the average, and unknown parameters can be replaced by their estimated values.

1.3.4 Numerical Computation of M-estimates

The calculation of M-estimates, as discussed in the last three sections, requires numerical solutions to optimization or root-finding problems. In this section we will derive these using the iterative reweighting method. The system of simultaneous location and dispersion introduced in Section 1.3.3 is discussed in this section. The special cases of univariate location (Section 1.3.1) and scale estimates (Section 1.3.2) can easily be simplified from the implemented algorithms by considering another parameter as the known value. Note that in practice both of the parameters are unknown. The role of numerical methods is vital in nonlinear regression parameter estimation and software design because nonlinear models have to be approximated by linear expansion and the performing of multiple numerical iterations.

The computation of location parameters from the simultaneous equations (1.21) depends on the value of the dispersion parameter. This can be computed before the location, or a simultaneous algorithm can be used to compute both. However, it is important to note that in numerical procedures, the initial estimate of the scale parameter will break the robustness of the location estimates. This is critical in nonlinear regression, which will be discussed in Chapter 3.

In general, consider an estimating iterative procedure, with values of location and dispersion in the kth iteration $(\hat{\mu}_k, \hat{\sigma}_k)$, and starting from initial values where $k = 1$. Let $\hat{t}_i = ((x_i - \hat{\mu})/\hat{\sigma})$. Depending on the known value of the dispersion we can solve three problems.

Problem 1: Location estimation with previously estimated dispersion: Assuming that the dispersion estimate computed previously is $(\hat{\sigma})$, the location estimate is the answer to the estimating equation (1.21a) for a fixed known value $(\hat{\sigma})$. Note that we can write Equation 1.21a as:

$$\frac{1}{\hat{\sigma}} \sum_{i=1}^{n} \frac{\psi_0(\hat{t}_i)}{t_i}(x_i - \hat{\mu}) = 0 \qquad (1.23)$$

Algorithm 1.1 Location with previously computed variance.

Step 1: Compute $\hat{\sigma} = \text{MADN}(x)$ and $\mu_0 = \text{Median}(x)$.
Step 2: For $k = 0, 1, \ldots$ compute the weights $w_{k,j}$ and then μ_{k+1} in (1.24).
Step 3: Stop iteration when $|\hat{\mu}_{k+1} - \hat{\mu}_k| < \epsilon\hat{\sigma}$.

Starting from an initial value for the location, the new iterative value can be computed as:

$$\hat{\mu}_{k+1} = \frac{\sum_{i=1}^n w_{k,i} x_i}{\sum_{i=1}^n w_{k,i}} \tag{1.24}$$

where $w_{k,i} = W_1((x_i - \hat{\mu}_k/\hat{\sigma})$ is the weight at iteration k for point x_i and

$$W_1(t) = \begin{cases} \psi(t)/t & t \neq 0 \\ \psi'(0) & t = 0 \end{cases} \tag{1.25}$$

If $W_1(t)$ is bounded and nonincreasing for $t > 0$, then the sequence converges to the solution of (1.21a), hence the algorithm for estimating the location to achieve a tolerance ϵ as the precision of the parameter can be defined:

Problem 2: Scale estimate: Suppose the location is a fixed known value, then the estimating equation of scale (1.21b) is equivalent to (1.17). It can be rewritten in the weighted form:

$$\frac{1}{n} \sum_{i=1}^n \frac{\rho_{\text{scale}}(t_i)}{t_i^2} \left(\frac{x_i - \hat{\mu}}{\hat{\sigma}} \right)^2 = \delta$$

An iterative solution is then:

$$\hat{\sigma}_{k+1}^2 = \frac{1}{n\delta} \sum_{i=1}^n w_{k,i}(x_i - \mu)^2 \tag{1.26}$$

where $w_{k,i} = W_2((x_i - \hat{\mu}_k/\hat{\sigma})$ is the weight at iteration k for point x_i and

$$W_2(t) = \begin{cases} \rho(t)/t^2 & t \neq 0 \\ \rho''(0) & t = 0 \end{cases} \tag{1.27}$$

This formula can be used to derive the iterative reweighting for estimating location and dispersion simultaneously. Meanwhile, for estimating the univariate scale estimate, without loss of generality, we can replace the zero value for location. If $W_2(t)$ is a bounded, even, continuous and nonincreasing function for $t > 0$, the sequence σ_k converges to the solution of (1.17), so the algorithm is as 1.2:

Problem 3: Simultaneous location and dispersion estimate: The simultaneous location and dispersion estimate in (1.21) can be derived from two old algorithms.

Algorithm 1.2 Location with previously computed variance.

Step 1: For $k = 0, 1, \ldots$ compute the weights $w_2 k, j$ and then σ_{k+1} in (1.26).
Step 2: Stop iteration when $|\hat{\sigma}_{k+1}/\hat{\sigma}_k| < \epsilon$.

Algorithm 1.3 Simultaneous location and dispersion estimate.

Step 1: Compute the starting values $\hat{\mu}_0, \hat{\sigma}_0$,

$$t_{k,i} = \frac{x_i - \hat{\mu}_k}{\hat{\sigma}_k}, \ i = 1, \ldots, n$$

and weights $W_1(t_{k,i})$ and $W_2(t_{k,i})$ from (1.25) and (1.27), respectively.
Step 2: Given μ_k, σ_k renews the iteration by

$$\hat{\mu}_{k+1} = \frac{\sum_{i=1}^{n} W_1(t_{k,i})x_i}{\sum_{i=1}^{n} W_1(t_{k,i})}, \quad \hat{\sigma}_{k+1} = \frac{\hat{\sigma}_k^2}{n\delta} \sum_{i=1}^{n} W_2(t_{k,i})t_{k,i}^2$$

1.4 Redescending M-estimates

Less informative distributions tend to have an exponential tail; that is, they might be too narrow and not wide enough to solve the outlier problem in practice. Better performance can be obtained with a long tailed distribution, and this may even increase the maximum risk slightly beyond its minimax value (Huber 1981). Consider the M-estimate subject to the side condition:

$$\psi(t) = 0 \text{ for } |t| > c \tag{1.28}$$

for arbitrary c. Figure 1.2a shows the ρ and ψ functions for the Huber, Hampel, Tukey bi-square and Andrew functions (see Table A.11 for definitions). The Huber rho function is unbounded while the Hampel, bi-square and Andrew functions are bounded and, when redescending, satisfy (1.28). Redescending ψ functions are beneficial when there are extreme outliers, but the improvement is relatively minor and their sensitivity increases to an incorrect scale value and the minimum of objective function $\sum \rho(x_i - \mu_n)$ can be a trapped, local minimum. Removing impossible outliers by careful data screening based on physical knowledge might be more effective and reduce the risk.

1.5 Breakdown Point

Intuitively, the breakdown point of an estimate is defined as the maximum proportion of arbitrarily large outliers (so, mathematically, with infinite values) that can be tolerated. In other words, it can be expressed as the smallest fraction of outliers that would cause the estimator to have an improperly large value. It

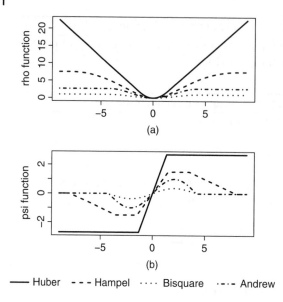

Figure 1.2 (a) ρ functions; (b) ψ functions.

—— Huber   - - - Hampel   ···· Bisquare   ·-·- Andrew

is known that the breakdown point concept is more useful for small-sample situations, so a finite sample breakdown point should be defined.

Definition 1.8   Finite sample breakdown point.

Let $X = (x_1, \ldots, x_n)$ be a fixed sample of size n. Huber (1981, Ch. 11.2) defined three ways of corrupting a sample:

(i) ε *contamination*: The sample is adjoined with m additional values $Y = (y_1, \ldots, y_m)$. Thus the fraction of "bad" values in the corrupted sample $Z = X \cup Y$ is $\varepsilon = m/(m+n)$.

(ii) ε *replacement*: A subset of size m is replaced by arbitrary values $(y_1, \ldots, y_m)$, hence the fraction of "bad" values in the corrupted sample Z is $\varepsilon = m/n$.

(iii) ε *modification*: Let π be an arbitrary distance function defined in the space of empirical measures. Let F_n be an empirical measure of sample X and G_m an empirical measure of sample Z with size m, such that $\pi(F_n, G_m) \leq \varepsilon$.

Let T_n be an estimator on the same Euclidean space for estimating parameter θ, $T(X)$ its value for sample X, and $T(Z)$ the value calculated for the ε modified/replaced/contaminated sample Z. The maximum bias caused by ε corruption is:

$$b(\varepsilon; X, T) = \sup_Z \| T(Z) - T(X) \| \tag{1.29}$$

where the supremum is taken over the set of all ε-corrupted samples Z. The breakdown point is then defined as:

$$\varepsilon^*(X, T) = \inf_{\varepsilon}\{\varepsilon|b(\varepsilon; X, T) = \infty\} \qquad (1.30)$$

which is a function of the original sample X and statistics T. Unless specified otherwise, we will work here with ε contamination.

For example, for estimating the location, say the mean is affected by a single outlier and has breakdown point zero. The median has breakdown $\frac{1}{2}$, and this is the highest possible breakdown point. A breakdown point of more than 50% is unreasonable because then it would appear that the good data have changed into bad data. Huber (1984) proved that the location M-estimator under regularity conditions has a breakdown point of 50%, or negligibly lower. If ψ is monotone, bounded and $\psi(-\infty) = -\psi(\infty)$, the breakdown point for the M-estimate (1.21a) is $\varepsilon^* = \frac{1}{2}$. The same is true if the scale parameter is unknown but estimated by a high breakdown point estimator "MAD", that is,

$$\hat{\mu}_M = \arg\min_{\mu} \sum_{i=1}^{n} \rho\left(\frac{x_i - \mu}{\text{MAD}}\right)$$

Huber (1984) proved the following theorem for redescending M-estimates with bounded ρ.

Theorem 1.9 Let min ρ occur at 0, $\rho(0) = -1$, ρ be a monotonic function increasing toward both sides, and $\lim \rho(0)$. If we put

$$\sum_{x_i \in X} \rho(x_i - T(X)) = -A$$

then the ε-contamination breakdown point of T is

$$\varepsilon^* = \frac{m^*}{n + m^*}$$

where m^* is an integer satisfying $\lceil A \rceil \leq m^* \leq \lfloor A \rfloor$. If $c < \infty$ such that $\rho(x) = 0$, for $|x| \geq c$ we have $m^* = \lceil A \rceil$.

Note that the breakdown point calculated using the above theorem depends on the ψ function and the data itself. For a redescending M-estimate and unbounded ρ, the following argument is satisfied.

Theorem 1.10 Assume ρ to be symmetric, $\rho(0) = 0$, and ρ increasing on both sides:

$$\lim_{|t| \to \infty} \rho(t) = \infty$$

but suppose

$$\lim_{|t| \to \infty} \frac{\rho(t)}{|t|} = 0$$

Moreover, assume $\psi = \rho'$ to be continuous, and there exists t_0 such that ψ is weakly increasing in $0 < t < t_0$ and weakly decreasing in $t_0 < t < \infty$. The breakdown point of the ε contamination for the M-estimate is then $\varepsilon^* = \frac{1}{2}$.

Remark 1.11 If ψ is nonmonotone and redescends to 0, as defined in Section 1.4, and if ρ is unbounded, then the breakdown point of the redescending M-estimate is high, with $\varepsilon^* = \frac{1}{2}$.

If ρ is bounded, the breakdown point is less than 0.5 but negligible.

1.6 Linear Regression

Multivariate linear regression is an intuitive approach to nonlinear inference, using a linear approximation of a nonlinear model (Chapter 2). Robust methods for location can be extended to linear regression and subsequently to nonlinear regression, and outlier detection methods for linear regression can similarly be extended to identification of outliers in nonlinear regressions. In the rest of this chapter, the theories of linear regression, robust linear regression, and outlier detection will be discussed.

Assume n couples of a random sample $z_i = (x_i, y_i)$, $i = 1, \ldots, n$ are observed, where x_i is an independent (predictor) variable and y_i is a dependent (response) variable. The linear regression model is defined as:

$$y_i = \beta_0 + \beta_1 x_i + \varepsilon_i, \; i = 1, \ldots, n \tag{1.31}$$

where (β_0, β_1) are unknown parameters, ε_i are errors, which in classical statistics are assumed to be independent with mean $E(\varepsilon_i) = 0$, and to have constant variance $Var(\varepsilon_i) = \sigma^2$. If there is a normal distribution, the MLE of parameter estimates can be employed. A more general linear model can be defined for when n observations contain p independent variables $(x_{i1}, \ldots, x_{ip}, y_i)$ and p unknown parameters $\beta_1, \ldots, \beta_p$:

$$y_i = \sum_{j=1}^{p} \beta_i x_i + \varepsilon_i, \; i = 1, \ldots, n \tag{1.32}$$

$$= \beta' \mathbf{x}_i + \varepsilon_i \tag{1.33}$$

The multivariate regression model can be written in a simple matrix form. Write the unknown parameters in a column vector $\beta = [\beta_1, \ldots, \beta_p]'$. The ith predictor observation with coordinates $\mathbf{x}_i = [x_{i1}, \ldots, x_{ip}]'$ is inserted in the columns of a design matrix $\mathbf{X}_{n \times p} = [\mathbf{x}_1, \ldots, \mathbf{x}_n]'$, the response values in a column vector $Y = [y_1, \ldots, y_n]$, and the residuals in an error vector $\varepsilon = [\varepsilon_1, \ldots, \varepsilon_n]$. Finally the multivariate linear regression model can be written in matrix form as:

$$Y = X\beta + \varepsilon \tag{1.34}$$
$$= \mu(\beta) + \varepsilon \tag{1.35}$$

Constant terms can be readily added to this formulation, and ignoring them in our inference here does not reduce the generality. Let $z_i = (y_i, x_i), i = 1, \ldots, n$ be i.i.d. observations of the linear regression. Let $G_0(x)$ be the distribution of predictor x_i and F_0 be the distribution of the error. Then the distribution of z_i is given by

$$H_0(z) = G_0(x)F_0(y - \beta_0' x_i) \tag{1.36}$$

where β_0 is the true value of parameter β Similarly, the true value of σ is denoted by σ_0. Fitted values, defined as $\hat{y}_i = x_i \hat{\beta}$, and residuals, defined as $r_i = y_i - \hat{y}_i$, can be represented in matrix form too:

$$\hat{Y} = X\hat{\beta}, r = Y - \hat{Y} \tag{1.37}$$

The least squares method is popular with scientists because of its ease of use. The sum of square errors (SSE) is defined as:

$$SSE(\beta) = \sum_{i=1}^{n}(r_i^2)$$
$$= (Y - \hat{Y})'(Y - \hat{Y})$$

so the least squares estimate of the parameters is the answer to the optimizing sum of squares, which requires the following normal equations to be solved:

$$(X'X)\hat{\beta} = X'Y \tag{1.38}$$

Therefore:

$$\hat{\beta}_{LS} = \arg\min_{\beta} SSE(\beta) \tag{1.39}$$
$$= (X'X)^{-1}X'Y \tag{1.40}$$

If the errors ε_i are i.i.d. with mean zero and constant variance σ^2, the covariance matrix of the error vector is $Cov(\varepsilon) = \sigma^2 I$, where I is an $n \times n$ identity matrix, and we have the following identities:

$$E(Y) = X\beta$$

which means the linear regression problem (1.34) is a more general form of the location model (1.2). Therefore a robust M-estimate can be defined. Table 1.1 shows a summary of formulas in linear regression. These can be used to develop intuitive formulas for nonlinear regression by replacing the design matrix in linear regression with the gradient of the nonlinear function.

Table 1.1 Linear regression formulas.

Statistic	Formula	Statistic	Formula
Location model	$\mu(\beta) = \mathbf{X}\beta$ $\hat{\mu}(\hat{\beta}) = \mathbf{X}\hat{\beta}$	Unbiasedness and limit distribution	$E(\hat{\beta}_{LS}) = \beta \; Var(\hat{\beta}) = \sigma^2(\mathbf{X'X})^{-1} \; \hat{\beta} \sim N_p(0, \sigma^2(\mathbf{X'X})^{-1})$
Parameter estimate	$\hat{\beta}_{LS} = (\mathbf{X'X})^{-1}\mathbf{X'Y}$	Variance	$\hat{\sigma}^2 = \frac{SSE(\hat{\beta})}{n-p}$
Prediction	$\hat{\mathbf{Y}} = \mathbf{X}\hat{\beta} = \mathbf{X}(\mathbf{X'X})^{-1}\mathbf{X'Y}$	SS-estimate	$SS(\hat{\beta}) = \mathbf{Y'}(I - H)\mathbf{Y}$
SSE	$SSE(\hat{\beta}) = \|\mathbf{Y} - \mathbf{X}\hat{\beta}\|$	Residuals	$\hat{\varepsilon} = \mathbf{Y} - \mathbf{X}\hat{\beta} \; E(\hat{\varepsilon}) = 0 \; Var(\hat{\varepsilon}) = \sigma^2(I - H)$
Hat matrix	$H = \mathbf{X}(\mathbf{X'X})^{-1}\mathbf{X'}$	$100(1-\alpha)\%$ Prediction interval	$\hat{\mathbf{Y}}_0 \in (\mathbf{X}_0'\hat{\beta} \pm s\sqrt{\mathbf{X'(X'X)^{-1}X}_0} t_{(n-p,\alpha/2)})$
Predictor covariance matrix	$Var(\hat{\mathbf{Y}}) = \sigma^2 H$		
$100(1-\alpha)\%$		statistics	$(\hat{\beta} - \beta)'\mathbf{X'X}(\hat{\beta} - \beta)/\sigma^2 \sim \chi_p^2$
Parameter CI	$\{\beta : (\beta - \hat{\beta})'\mathbf{X'X}(\beta - \hat{\beta}) \le pF_{v,n-p}^\alpha\}$		$(\hat{\beta} - \beta)'\mathbf{X'X}(\hat{\beta} - \beta)/ps^2 \sim F_{(p,n-p)}$
R square	$R^2 = SSR/SSTO$	Error SS	$SSE = (\mathbf{Y} - \hat{\mathbf{Y}})'(\mathbf{Y} - \hat{\mathbf{Y}})$
Regression SS	$SSR = (\hat{\mathbf{Y}} - \overline{\mathbf{Y}})'(\hat{\mathbf{Y}} - \overline{\mathbf{Y}})$		$MSE = SSE/(n - p - 1)$
	$MSR = SSR/(p)$		
Total SS	$SSTO = (\mathbf{Y} - \overline{\mathbf{Y}})'(\mathbf{Y} - \overline{\mathbf{Y}})$	F ratio	$F = MSR/MSE \sim F(p, n - p - 1)$
	$MSTO = SSTO/(n - 1)$		

SS, sum of squares.

1.7 The Robust Approach in Linear Regression

There are several robust methods for estimating the location and scale parameters in a linear regression. This section discuss the MM-estimate, which was defined by Yohai (1987). The logical path of the section is to end up with a simultaneous estimate of the location and scale parameters – the MM-estimate – which is a high breakdown point (HBP) estimate with asymptotic normality and strong consistency.

Simple linear regression (1.31) and general multiple linear regression (1.34), can both be considered as types of location and scale model (1.18) with multiple parameters. Consequently a robust M-estimate can be defined.

Analogous to Section 1.3.3, if the scale parameter σ is known, the location M-estimate is defined in the following way.

Let ρ be a real function satisfying the following assumptions.

A 1.12

(i) $\rho(0) = 0$
(ii) $\rho(-t) = \rho(t)$
(iii) $0 \leq u < v$ implies $\rho(u) \leq \rho(v)$
(iv) ρ is continuous
(v) let $a = \sup \rho(t)$ then $0 < a < \infty$
(vi) let $\rho(t) < a$ and $0 \leq u < v$ then $\rho(u) < \rho(v)$.

Given a sample of size n $(y_i, \mathbf{x}_i)$, $i = 1, \ldots, n$ and residuals defined as $r_i(\beta) = y_i - \beta' \mathbf{x}_i$, the M-estimate of location parameter β, for known variance σ^2, is the solution of the optimization problem:

$$\hat{\mu} = \arg \min_{\mu} \sum_{i=1}^{n} \rho \left(\frac{r_i(\beta)}{\sigma} \right)$$

When the variance is unknown and the location has been previously calculated or is known, the M-estimate of scale $\hat{\sigma}$ is defined as the answer to the equation:

$$\frac{1}{n} \sum \rho \left(\frac{r_i(\beta)}{\hat{\sigma}} \right) = b$$

where b is a constant, which may be defined by

$$E_{\phi}(\rho(t)) = b$$

We have to estimate the location and scale parameters simultaneously. In order to achieve efficiency and an HBP estimate, the MM-estimate algorithm for both location and scale parameters is defined as shown in Algorithm 1.4.

Algorithm 1.4 Robust M-estimate for linear regression

Stage 1: Take an estimate $\hat{\beta}_{0,n}$ of β with high breakdown point, possibly 0.5.

Stage 2: Compute the residuals $r_i(\hat{\beta}_{0,n}) = y_i - \mathbf{x}_i \hat{\beta}_{0,n}$.

Stage 3: Compute the M-estimate of scale $\hat{\sigma}_n = \sigma(r_i(\hat{\beta}_{0,n}))$ using a function ρ_0 obtained from

$$\frac{1}{n} \sum_{i=1}^{n} \rho_0 \left(\frac{r_i(\hat{\beta}_{0,n})}{\hat{\sigma}_n} \right) = b, \tag{1.41}$$

$b = E_\phi(\rho_0(t))$. Define the constant a such that:

A 1.13 $b/a = 0.5$

where $a = \max \rho_0(t)$. As Huber proves, this implies that this scale estimate is (50%) HBP.

Stage 4: Let ρ_1 be another function satisfying regularity conditions

A 1.14 $\rho_1(t) \leq \rho_0(t)$

and

A 1.15 $\sup \rho_1(t) = \sup \rho_0(t) = a$

Let $\psi_1 = \rho_1'$. Then the MM-estimate $\hat{\beta}_{1,n}$ is defined as the solution of

$$\sum_{i=1}^{n} \psi_1 \left(r_i(\beta)/\hat{\sigma}_n \right) \mathbf{x}_i = 0 \tag{1.42}$$

which verifies

$$S_n(\hat{\beta}_{1,n}) < S_n(\hat{\beta}_{0,n}) \tag{1.43}$$

where

$$S_n(\beta) = \sum_{i=1}^{n} \rho_1(r_i(\beta)/\hat{\sigma}_n)$$

and $\rho_1(0/0)$ is defined as 0.

The MM-estimate was defined by Yohai (1987), who proved that the estimate is HBP, asymptotic normal, and strongly consistent. This method is used in other parts of the book to develop estimators for nonlinear regression.

Definition 1.16 Analogous to ε contamination (Definition 1.8) we can define the contamination and breakdown point for linear regression. Let $\mathbf{Z}_n = \{(y_i, \mathbf{x}_i), i = 1, \ldots, n\}$ be a set of n observations. The corrupted sample, with m additional values $\mathbf{W}_m$ where $\mathbf{Y}_n \cup \mathbf{W}_m$, and with sample of size $n + m$, contains observations of both samples. Let $\mathbf{T}_n$ be the estimate corresponding to a sample of size n. The bias can be written as

$$b(m, \mathbf{T}, \mathbf{Z}_n) = \sup_Z \| \mathbf{T}_{m+n}(\mathbf{Z}_n \cup \mathbf{W}_m) - \mathbf{T}_n(\mathbf{Z}_n) \|$$

and the finite sample breakdown point (1.30) is

$$\varepsilon^*(Z_n, T) = \inf\{m(m+n) | b(m, T, Z_n) = \infty\}$$

Yohai (1987) proved the following theorems (we omit the proofs, which may be found in Yohai's papers (1987; 1985)):

Theorem 1.17 For ε contamination of size m, define c_n as

$$c_n = \max_{\theta \in \mathfrak{R}^p} \#\{i : 1 \le i \le n \text{ and } \beta' X_i = 0\}/n \tag{1.44}$$

Suppose assumptions A 1.12–A 1.15 are satisfied. Then, for given $\varepsilon < (1 - 2c_n)/(2 - 2c_n)$ and k_0, there exists a k_1 such that $m/n + m \le \varepsilon$, and $S_{m+n} \le k_0$ implies:

$$\inf_{\|\theta\| \ge k_1} \sum_{i=1}^{m+n} \rho_1(r_i(\theta)/S_{m+n}) > \sum_{i=1}^{m+n} \rho_1(r_i(\hat{\theta}_{0,m+n})/S_{m+n})$$

for all samples $Z_n \cup W_m$.

Note that Theorem 1.17 implies that the absolute minimum of $S_n(\beta)$ exists and obviously satisfies (1.42) and (1.43). However, any other value of β that satisfies (1.42) and (1.43) is a local minimum and is an MM-estimate with HBP and high efficiency.

Theorem 1.18 Suppose ρ_0 and ρ_1 satisfy assumptions A1.12–1.15 and $c_n < 0.5$. Then, if $\hat{\beta}_0 = \{\hat{\beta}_{0,n}\}_{n \ge p}$ is any sequence of estimates that satisfies (1.43):

$$\varepsilon^*(\hat{\beta}_1, Z_n) \ge \min(\varepsilon^*(\hat{\beta}_0, Z_n), (1 - 2c_n)/(2 - 2c_n)) \tag{1.45}$$

Note that $c_n \to 0$ as $n \to \infty$, so $(1 - 2c_n)/(2 - 2c_n) \to 0$. Therefore, the above theorem implies that if $\varepsilon^*(\hat{\beta}_0, Z_n)$ is asymptotically 0.5, $\varepsilon^*(\hat{\beta}_1, Z_n)$ is asymptotically 0.5.

Theorem 1.19 Let i.i.d. observations $z_i = (y_i, x_i)$ be given, with distribution G_0. Assume that ρ_0 satisfies A 1.12 and $\{\hat{\beta}_{0,n}\}_{n \ge p}$ is a sequence of estimates that is strongly consistent for θ_0. Then the M-estimate of scale $(\hat{\sigma}_n)$ obtained from Stage 2 of Algorithm 1.4 (see (1.41)) is strongly consistent for σ_0.

The consistency of $\hat{\beta}_{1,n}$ requires the following two assumptions to hold:

A 1.20 The function $g(a) = E_{F_0}[\rho_1((u-a)/\sigma_0)]$, where σ_0 is defined by $E_{F_0}[\rho_0(u/\sigma_0)] = b$, has a unique minimum at $a = 0$.

A 1.21 $P_{G_0}(\beta' x = 0) < 0.5$ for all $\beta \in \mathfrak{R}^p$.

If ρ_1 satisfies A 1.12, then the sufficient condition for 1.20 is given as below:

A 1.22 The error distribution F_0 has density f_0 with the following properties:

(i) f_0 is even

(ii) $f_0(u)$ is monotone nonincreasing in $|u|$

(iii) $f_0(u)$ is strictly decreasing in $|u|$ in the neighborhood of 0.

Theorem 1.23 Let i.i.d. observations $z_i = (y_i, x_i)$ be given with distribution G_0. Assume ρ_0 and ρ_1 satisfy assumptions A 1.12–A 1.21. Assume also that sequence $\{\hat{\beta}_{0,n}\}_{n \geq p}$ is strongly consistent for θ_0. Then any other sequence $\{\hat{\beta}_{1,n}\}_{n \geq p}$ for which $S_n(\hat{\beta}_{1,n}) < S_n(\hat{\beta}_{0,n})$ (Equation 1.43) is strongly consistent too.

Asymptotic normality of the M-estimate requires four of the x_is, but the MM-estimate requires a second moment. We need some additional assumptions:

A 1.24 ρ_1 is an even, twice continuously differentiable function and there exists m such that $|u| \geq m$ implies $\rho_1(u) = a$.

A 1.25 G_0 has second moments and $V = E_{G_0}(x_i x_i')$ is nonsingular.

Theorem 1.26 Let $z_i, i = 1, \ldots, n$ be i.i.d. with distribution H_0. Assume ρ_1 satisfies 1.24 and G_0 satisfies 1.25. Let $\hat{\sigma}_n$ be an estimate of error scale that converges strongly to σ_0. Let $\hat{\beta}_n$ be a sequence of estimates that satisfies (1.42) and which is strongly consistent to the true value β_0. Then

$$n^{1/2}(\hat{\beta}_n - \beta_0) \to_d N(0, \sigma_0^2 [A(\psi_1, F_0)/B^2(\psi_1, F_0)] V^{-1}) \tag{1.46}$$

where

$$A(\psi, F) = E_F(\psi^2(u/\sigma_0))$$

and

$$B(\psi, F) = E_F(\psi'(u/\sigma_0))$$

One way to choose ρ_0 and ρ_1 satisfying A 1.12, A 1.14 and A 1.15 is as follows. Let ρ satisfy A.1.14, and let $0 < k_0 < k_1$. Let $\rho_0(t) = \rho(t/k_0)$ and $\rho_1(t) = \rho(t/k_1)$. Then Assumption 1.12*iii* implies $\rho(t/k_1) < \rho(t/k_0)$.

The scale estimate $\hat{\sigma}_n$ is the answer to the equation

$$\frac{1}{n} \sum_{i=1}^{n} \rho\left(\frac{r_i(\hat{\beta}_{0,n})}{k_0 \hat{\sigma}_n}\right) = b$$

The value of k_0 should be chosen such that $b/a = 0.5$, $b = E_{F_0}(\rho_1(t))$, and $a = \max \rho_0(t)$. Thus, from Theorem 1.18, the MM-estimate $\hat{\beta}_{1,n}$ is HBP.

The value of k_1 can be chosen to determine the asymptotic efficiency. The MM-estimate of β in Stage 3 of the MM-estimate algorithm is computed as

$$\hat{\beta}_{1,n} = \arg \min_{\beta} \sum_{i=1}^{n} \rho \left(\frac{r_i(\beta)}{k_1 \hat{\sigma}_n} \right)$$

or equivalently by the equation

$$\sum_{i=1}^{n} \psi \left(\frac{r_i(\beta)}{k_1 \hat{\sigma}_n} \right) \mathbf{x}_i = 0$$

From the asymptotic normal distribution of the MM-estimate $\hat{\beta}_n$ (Equation 1.46), if the errors have an $N(0,1)$ distribution, $\sigma_0 = 1$ and the variance depends only on k_1 and not k_0 explicitly. To achieve a certain amount of efficiency, for example 95%, the Fisher information for linear regression is $I(\beta) = (1/\sigma_0^2)\mathbf{X}'\mathbf{X}$ and the efficiency of the MM-estimate is equal to

$$
\begin{aligned}
\textit{eff}(\hat{\beta}_{1,n}) &= \frac{1/I(\beta)}{Var(\hat{\beta}_{1,n})} \\
&= \frac{\sigma_0^2(\mathbf{X}'\mathbf{X})^{-1}}{\sigma_0^2[A(\psi_1, F_0)/B^2(\psi_1, F_0)]V^{-1}} \\
&= \frac{B^2(\psi_1, F_0)}{A(\psi_1, F_0)}
\end{aligned}
$$

1.8 S-estimator

S-estimate stands for scale estimate, which is the equivalent of the M-estimator for the scale parameter σ. It is used in defining the τ-estimate. The S-estimate was defined by Rousseeuw and Yohai (1984) for linear regression, and they proved that it had HBP and asymptotic normal properties. The S-estimate is difficult to compute, although good algorithms have been proposed. The importance of the S-estimate is in helping to compute tuning constants in methods for obtaining HBP estimators.

Define the dispersion function $S(r_1, \ldots, r_n)$ to be a function of any sample $(r_1, \ldots, r_n)$ of real numbers. It is the solution of the Huber equation of the form:

$$\sum_{i=1}^{n} \rho \left(\frac{r_i(\beta)}{S} \right) = nk \tag{1.47}$$

where $k = E_\phi(\rho)$ is the mean of the real valued function ρ and ϕ is the standard normal density function. The real valued function ρ was defined in Chapter 1, and for the S-estimate we assume it satisfies the following assumptions:

A 1.27 ρ is symmetric and continuously differentiable and $\rho(0) = 0$.

A 1.28 There exists $c > 0$ such that ρ is strictly increasing on $[0, c]$ and constant on $[c, \infty)$.

We want to estimate the function model parameter β and scale σ. For each vector β, we obtain residuals $r_i(\beta) = y_i - \mathbf{x}_i \beta$. The dispersion function $S(r_1(\beta), \ldots, r_n(\beta))$ is then calculated from Equation 3.7. After that, we define the S-estimate of β (denoted by $\hat{\theta}_S$) as:

$$\hat{\beta}_{n,S} = \arg\min_{\beta} S(r_1(\beta), \ldots, r_n(\beta)) \tag{1.48}$$

and the final scale estimator is

$$\hat{\sigma}_{n,S} = S(r_1(\hat{\beta}_S), \ldots, r_n(\hat{\beta}_S)) \tag{1.49}$$

S-estimators are affine equivariant, but their calculation is difficult.

Rousseeuw and Yohai (1984) studied the breakdown property of the S-estimate and proved that under certain conditions it is consistent and asymptotically normal (see the following theorems, but the proofs are omitted).

Theorem 1.29 Breakdown point of the S-estimate. Assume:

A 1.30 $\frac{E_\phi(\rho)}{\rho(c)} = \frac{1}{2}$, where c is the consistency factor.

Under A 1.27–A 1.30, the solution (1.48) exists and the breakdown of the S-estimate is equal to

$$\varepsilon^* = ([\frac{n}{2}] - p + 2)/n$$

which tends to 0.5 as $n \to \inf$. If condition 1.30 is replaced by $\frac{E_\phi(\rho)}{\rho(c)} = \lambda$, where $0 < \lambda < 1/2$, then the corresponding S-estimate has a breakdown point tending to λ.

For the proofs, see Rousseeuw and Yohai (1984).

Theorem 1.31 *Consistency* Let ρ be a function satisfying 1.27 and 1.28, with derivative $\rho' = \psi$. Assume that

(i) $\psi(t)/t$ is nonincreasing for $t > 0$
(ii) $E(\|X\|) < \infty$.

Let $\hat{\beta}_{n,S}$ and $\hat{\sigma}_{n,S}$ be solutions of (1.48) and (1.49), respectively. Then:

$$\hat{\beta}_{n,S} \to \beta_0, a.s. \tag{1.50}$$

$$\hat{\sigma}_{n,S} \to \sigma_0, a.s. \tag{1.51}$$

where β_0 and σ_0 are the actual values of the parameters. For the proofs, see Maronna and Yohai (1981, Thm 2.2 and 3.1). Note that the S-estimate satisfies the same first-order necessary conditions as M-estimates.

Theorem 1.32   Asymptotic normality.
   Let $\beta_0 = 0$ and $\sigma_0 = 0$ for simplicity. If the conditions of Theorem 1.31 hold and also:

(iii) ψ is differentiable in all but a finite number of points, $|\psi'|$ is bounded and $\psi' d\phi > 0$

(iv) $E[XX^T]$ is nonsingular and $E[\|X\|^3] < \infty$

then

$$n^{1/2}(\hat{\beta}_{n,S} - \beta_0) \to N\left(\mathbf{0}, E[XX^T]^{-1}\frac{\int \psi^2 d\phi}{(\int \psi' d\phi)^2}\right) \tag{1.52}$$

$$n^{1/2}(\hat{\sigma}_{n,S} - \sigma_0) \to N\left(\mathbf{0}, E[XX^T]^{-1}\frac{\int (\rho(y) - k)^2 d\phi(y)}{(\int y\psi(y)d\phi(y))^2}\right) \tag{1.53}$$

The asymptotic distributions (1.52) and (1.53) are useful for practical problem inferences so they help us to compute the covariance matrix of parameter estimates. In particular, in Section 3.9 we extend them to nonlinear regression.

1.9   Least Absolute and Quantile Esimates

About a century ago, Edgeworth (1887) observed that methods of estimation based on minimizing sums of absolute residuals could be far superior to least squares methods under non-Gaussian error conditions. However, computation remained a major problem until simple linear programming algorithms were developed by researchers such as Wagner (1959) and Charnes et al. (1955), and algorithms for the l_1 norm were provided by Barrodale and Roberts (1974), Bartels and Conn (1980), and others. These algorithms are readily extended to linear quantile regression, as introduced by Koenker and Bassett (1978) and Koenker and D'Orey (1987), and can also be applied to nonlinear regression, as was done by Koenker and Park (1996).
   The least absolute (LA) value estimates, known as l_1 norm estimates, minimize the L_1 criterion, defined as

$$\hat{\theta}_{LA} = \arg\min_{\theta} \sum_{i=1}^{n} |r_i(\theta)| \tag{1.54}$$

where the residuals are defined as

$$r_i(\beta) = (y_i - X_i\beta), i = 1, \dots, n \tag{1.55}$$

The LA estimator of β may be formulated as a linear program. The primal for the l_1 linear program may be written as:

$$\max\{\mathbf{1}'_n\varepsilon^+ + \mathbf{1}'_n\varepsilon^- | (\beta, \varepsilon^+, \varepsilon^-) \in \mathfrak{R}^p \times \mathfrak{R}^{2n}_+, X\beta + \varepsilon^+ - \varepsilon^- = Y\} \tag{1.56}$$

where $\mathbf{1}_n$ is an n-vector of 1s, $\varepsilon_i^+ = max(\varepsilon_i, 0)$, and $\varepsilon_i^- = -min(\varepsilon_i, 0)$. The simplex or interior point method can be used to solve the linear programming problem (1.56). Having distinguished the positive and negative parts of the residual vector, we are simply minimizing a linear function subject to linear constraints. The dual problem may be written as

$$\max\{Y'\mathbf{d} | \mathbf{d} \in \Omega = d \in [-1, 1]^n, X'\mathbf{d} = 0\}$$

The dual variables, $\mathbf{d}$, may be viewed as Lagrange multipliers on the constraints; that is, the marginal costs of relaxing the constraints. If ε_i is nonzero, then $d_i = \text{sgn}(\varepsilon_i)$; otherwise, when $\varepsilon_i = 0, \mathbf{d}_i \in (-1, 1)$. By complementary slackness there will be, barring degeneracy in the primal problem, exactly p of the ε_is equal to zero at an optimum. Consequently, p of the $\mathbf{d}_i$s are not equal to ± 1. For more detail see Koenker and D'Orey (1987) and Koenker and Bassett (1978) for linear, and Kennedy and Gentle (1980) for nonlinear regression formulation and computation.

The quantile estimate can be applied in linear regression, a process known as linear quantile regression. Suppose that the αth conditional quantile function is $Q_{Y|X}(\alpha) = X\beta_\alpha$. Given the distribution function of Y, β_α can be obtained:

$$\beta_\alpha = \arg\min_\beta E[\rho(Y - X\beta)] \qquad \rho_\theta(u) = u(\alpha - I(u < 0))$$

For an indicator function I and loss function ρ we have

$$\min_u E[\rho(Y - u)] = \min_u (\alpha - 1) \int_{-\infty}^{u} (y - u)dF_Y(y) + \alpha \int_{u}^{\infty} (y - u)dF_Y(y)$$

which is the αth quantile of the distribution of Y. Replacing the sample analog, symmetric l_1 criterion, with an asymmetric linear criterion, gives an estimate of β

$$\ell_\alpha(\beta) = \sum_{i=1}^{n} \rho_\alpha[r_i(\beta)] \qquad \rho_\alpha(u) = y(\alpha - I(u < 0))$$

In this way, the quantile regression of Koenker and Bassett (1978) will be obtained. The dual problem is now

$$\max\{Y'\mathbf{d} | \mathbf{d} \in \Omega = d \in [\alpha - 1, \alpha]^n, X'\mathbf{d} = 0\}$$

1.10   Outlier Detection in Linear Regression

In this section, methods for outlier detection in linear regression are discussed. Some of these methods are used in Chapter 6 to identify outliers in nonlinear regressions.

There are several approaches to detecting outliers. There are statistics that measure the influences of different kinds of data points, or that measure the distance of the data points from the fitted regression line. Another approach is to isolate outlying data using robust methods. Since the robust fitted regression line is not swamped by outlier data, the computed residuals of the outliers can be large, revealing the large distance to the regression line.

Another popular method is to study what happens when a single observation or a group of observations is deleted.

Consider the general linear regression (1.34). Using the hat matrix, defined by $H = \mathbf{X}(\mathbf{X}'\mathbf{X})^{-1}\mathbf{X}'$, the prediction expression (see Table 1.1) can be written as:

$$\hat{\mathbf{Y}} = H\mathbf{Y}$$

$$\hat{y}_i = h_{ii}y_i + \sum_{j \neq i; j=1}^{n} h_{ij}y_j \tag{1.57}$$

In this equation, if h_{ii} is large relative to the remaining terms, the fitted value $\hat{y}_i$ is dominated by response $\hat{y}_j$, so h_{ij} is interpreted as the amount of influence or leverage of y_j on $\hat{y}_i$. Following this reasoning, Hoaglin and Welsch (1978) suggested direct use of h_{ij} as a diagnostic to identify high leverage points.

A common way of developing an influence detection method is to refit a model by deleting a special case or a set of cases. The amount of change of certain statistics – the parameter estimates, predicted likelihoods, residuals, and so on – can be observed for the measure when recalculated with the *i*th data point removed. The notation $(-i)$ is used for each removed observation. Subsequently, the new estimates can be utilized in the computation of the influence measures.

Some statistical measures for identifying outliers are briefly discussed in the following sections.

1.10.1   Studentized and Deletion Studentized Residuals

Studentized residuals (hereafter referred to as t_i) are used for identifying outliers. They are standardized by dividing the residuals by their standard error. Using the prediction form given by the hat matrix and the diagonal form of the covariance matrix of residuals (see Table 1.1), the standardized residuals can be written as:

$$t_i = \frac{r_i}{\sigma\sqrt{1 - h_{ii}}}. \tag{1.58}$$

Note that the variance σ^2 is unknown, so we replace it by its estimate $\hat{\sigma}$. The studentized residuals are then defined as

$$t_i = \frac{r_i}{\hat{\sigma}\sqrt{1 - h_{ii}}} \tag{1.59}$$

where h_{ii} is the diagonal of the hat matrix H. The deleted studentized residuals (d_i) are defined as

$$d_i = \frac{r_i}{\hat{\sigma}_{(-i)}\sqrt{1 - h_{ii}}}$$

where $\hat{\sigma}_{(-i)}$ is the estimated standard deviation in the absence of the ith observation. The residuals, denoted by $r_i = y_i - f(x_i; \hat{\theta})$, are obtained from the ordinary least squares-, M- or MM-estimates. The ith observation is considered an outlier if $|t_i|$ or $|d_i| > 2.5$ or 3 (Anscombe and Tukey 1963; Srikantan 1961)

1.10.2 Hadi Potential

Hadi (1992) proposed the Hadi potential, given by $p_i i$, to detect high leverage points or large residuals:

$$p_{ii} = \frac{h_{ii}}{1 - h_{ii}}$$

He proposed the cut-off point for $p_i i$ to be $\text{Median}(p_{ii}) + c \times \text{MADN}(p_{ii})$, where MADN represents the normalized mean absolute deviance, defined by:

$$\text{MADN}(p_{ii}) = \text{Median}\{p_{ii} - \text{Median}(p_{ii})\}/0.6745 \tag{1.60}$$

c is an appropriately chosen constant, such as 2 or 3.

1.10.3 Elliptic Norm (Cook Distance)

The Cook distance (denoted by CD), which was defined by Cook and Weisberg (1982), is used to assess influential observations. An observation is influential if the value of CD is greater than 1. They defined CD as:

$$CD_i = \frac{(\hat{\beta} - \hat{\beta}_{(-i)})^T (X^T X)(\hat{\beta} - \hat{\beta}_{(-i)})}{p\hat{\sigma}^2}$$

where $\hat{\beta}_{(-i)}$ is the parameter estimate when the ith observation is removed. It can be shown that the CD_i can be written as:

$$CD_i = \frac{t_i^2}{p} \frac{h_{ii}}{1 - h_{ii}}$$

since this form avoids deleting observations and is numerically more efficient, especially when we extend it to nonlinear regression. The cut-off point is equal to 1; that is, the expectation of a 50% confidence ellipsoid of parameter estimates.

1.10.4 Difference in Fits

Difference in fits (DFFITS) is another diagnostic parameter used when measuring influence and was defined by Belsley et al. (1980). For the ith observation, DFFITS is defined as

$$DFFITS_i = \left(\sqrt{\frac{h_{ii}}{1 - h_{ii}}} \right) |d_i|$$

Belsley et al. considered an observation to be an outlier when DFFITS exceeds a cut-off point of $2\sqrt{p/n}$.

1.10.5 Atkinson's Distance

The Atkinson distance (denoted by C_i for observation i) was developed by Atkinson (1981), who also studied its properties (Atkinson 1982; 1986). It is used to detect influential observations, defined as:

$$C_i = \left(\sqrt{\frac{n - p}{p} \frac{h_{ii}}{1 - h_{ii}}} \right) |d_i|$$

The cut-off point is suggested to be equal to 2.

1.10.6 DFBETAS

DFBETAS is a measure of how much an observation has affected an estimate of a regression coefficient (there is one DFBETA for each regression coefficient, including the intercept). For linear regression with design matrix $\mathbf{X}$, the DFBETAS for the ith data point and jth parameter can be computed as:

$$DFBETAS_{j(i)} = \frac{\hat{\beta}_j - \hat{\beta}_{j(i)}}{\sqrt{\hat{\sigma}^2_{(i)} (\mathbf{X}^T \mathbf{X})^{-1}_{jj}}} \tag{1.61}$$

For small/medium datasets, an absolute value of 1 or greater is 'suspicious'. For large datasets, absolute values larger than $2/\sqrt{n}$ are considered highly influential.

2

Nonlinear Models: Concepts and Parameter Estimation

2.1 Introduction

This chapter presents the basic concepts, notation, and theories of nonlinear regression that are used in this book. Nonlinear regression can be represented as a more general functional form of linear regression. Some models in nonlinear regression can therefore be approximated by linear expansions. There are, however, differences that should be taken into consideration. Although linear regression and linear models can be used to describe several statistical models, it is unreasonable to assume that linear models are sufficient. However, the theoretical and computational difficulties of nonlinear regression have resulted in its being neglected and sometimes forgotten by statisticians.

Powerful computational tools and efficient numerical algorithms have been developed for nonlinear modelling. In particular, several R packages for nonlinear regression have been developed, and these are discussed in this book. In general, estimation methods in nonlinear regression require the application of numerical algorithms, which are also explored in this book. Finally, numerical methods for robust nonlinear regression are explained, and an R package called `nlr`, which was developed by the author, is discussed.

In applied sciences, there are appropriate methods for constructing nonlinear models for a statistics data set in certain phenomena. For example, differential equation models are widely used in biostatistics and agriculture to find suitable nonlinear functions for a data set under known conditions. There are also efficient methods for selecting the best nonlinear model from a set of possible functions (Bunke et al. 1995a,b). However, it is not possible to set out a systematic approach to finding a nonlinear regression model for a statistics data set because, in practice, we are faced with an infinite number of possible nonlinear models that might fit the data. This problem, along with others – the initial values, the complexity of nonlinear models, the convergence problem,

Robust Nonlinear Regression: with Applications using R, First Edition.
Hossein Riazoshams, Habshah Midi, and Gebrenegus Ghilagaber.
© 2019 John Wiley & Sons Ltd. Published 2019 by John Wiley & Sons Ltd.
Companion website: www.wiley.com/go/riazoshams/robustnonlinearregression

curvature, and many other undefined issues – have caused nonlinear regression to be neglected.

2.2 Basic Concepts

Seber and Wild (2003) defined a nonlinear model as a general statistial relationship between a set of variables including at least one random term. We express the relationship via the function $y = f(x_1, x_2, \ldots, x_n)$, in which y is the response and $(x_1, x_2, \ldots, x_n)$ are the explanatory variable values, which are assumed to be known. Sometimes the mathematical form of f is known apart from some unknown constants or coefficients, known as parameters, for example a vector θ.

We can write the mathematical relation as $y = f(x_1, x_2, \ldots, x_n; \theta)$, where the function f is entirely known except for parameter vector $\theta \in \Theta$, which has to be estimated.

Parameters are relationships for a physical process that follows some accepted scientific laws. These physical processes are used by researchers and therefore, apart from special cases, cannot be fitted by linear transformations, partial linear physical processes, or smoothed curves using nonparametric methods like kernel fitting. Therefore, the parametric form of nonlinear functions still has to be taken into consideration. In practical examples, the values of parameter vector θ encapsulate a physical concept and represent a meaningful factor in scientific research.

Definition 2.1 A nonlinear model is the general mathematical model representing the relationship between the explanatory and the response variables with unknown parameters denoted by:

$$y_i = f(x_i; \theta) + \varepsilon_i, \ i = 1, \ldots, n \tag{2.1}$$

where y_i are the observable responses and x_i are k dimensional vectors whose values are assumed known. $\theta \in \Re^p$ is a p-dimensional unknown parameter vector, and ε_i are the errors satisfying some stochastic regularity conditions.

A more general matrix notation is used by Bates and Watts (2007). The model can be written as:

$$\mathbf{y} = \boldsymbol{\eta}(\theta) + \varepsilon \tag{2.2}$$

where $\mathbf{y} = [y_1, y_2, \ldots, y_n]^T$ is an $n \times 1$ response vector, $\boldsymbol{\eta}(\theta) = [f(x_1; \theta), \ldots, f(x_n; \theta)]^T$ is an $n \times 1$ vector of nonlinear model functions $f(x_i; \theta)$, $x_i \in \Re^q$ are predictors, and $\varepsilon = [\varepsilon_1, \varepsilon_2, \ldots, \varepsilon_n]^T$ is an $n \times 1$ vector of errors.

The random part in a nonlinear regression model is the error ε. If the distribution of the errors is known (say, the normal distribution), it can help us to make statistical inferences about the parameters $\boldsymbol{\theta}$ or the response **y**. Even if the distribution of errors is not known, there are several solutions for estimation and subsequent inference: the least squares estimator, robust statistics, and so on.

Traditionally, it is assumed that errors $\varepsilon_i \overset{iid}{\sim} D(0, \sigma), i = 1, \ldots, n$, are identically independently distributed (i.i.d.) as D (say, normal) with mean zero (denoted by $E(\varepsilon_i) = 0$) and equal variance (denoted by $Var(\varepsilon_i) = \sigma^2$). These assumptions may not hold true in real-life examples, but for null conditions there are appropriate methods, which are discussed in the book. An example of a nonlinear regression that follows these regularity assumptions is Lakes data, which also include the outlier points (Table A.4).

The constant error variance ($Var(\varepsilon_i) = \sigma^2$) is called the homogeneous variance. In contrast, when the variance is not constant, it is called *heteroscedastic* (or *heterogeneous*). Heterogeneity of variance is a widespread general concept. In Chapter 4, the details of heteroscedastic variance and several possible models to encapsulate it will be discussed. Heteroscedasticity of variance can be seen in natural phenomena. An example, in a model of chicken growth data will be covered in Figure 4.1.

Another natural situation that might break the classical assumption is autocorrelation of errors. Again, we will cover an example in later chapter, in the shape of a model of net money data for Iran (see Figure 5.2). Autocorrelation of errors is discussed in Chapter 5.

The derivatives of the nonlinear function model f might not exist in practice, but the asymptotic normality of least square errors requires the existence of derivatives. In the computation part of the book, if the derivatives exist, the iteration algorithms will be faster. If the first and second derivatives of expectation surface with respect to the parameters exist, then we have following definitions:

Definition 2.2 *Gradient vector* The gradient vector of the nonlinear function model $f(x_i; \theta)$, denoted by $\dot{\eta}_i(\theta)$ for $i = 1, \ldots, n$, is the vector with the elements of the first derivative of the nonlinear function model with respect to the parameter vector, defined as

$$\dot{\eta}_i^T(\theta) = \frac{\partial f(x_i; \theta)}{\partial \theta} = \left[\frac{\partial f(x_i; \theta)}{\partial \theta_1}, \ldots, \frac{\partial f(x_i; \theta)}{\partial \theta_p} \right]_{p \times 1}^T, i = 1, \ldots, n \quad (2.3)$$

We denote the derivative of f at point x_i with respect to the jth parameter θ_j by f'_{ij}. For computational purposes, the gradient can be stored in an $n \times p$ matrix, referred to as the gradient matrix, denoted by $\dot{\eta}$, which includes derivative $\dot{\eta}_i^T$ values in the rows of the matrix; that is,

$$\dot{\eta} = [f'_{ij}], i = 1, \ldots, n, j = 1, \ldots, p \quad (2.4)$$

Definition 2.3 *Hessian matrix* The Hessian matrix, denoted by $\ddot{\eta}_i(\theta)$, is the matrix of the second derivative of the nonlinear function model with respect to the parameter vector, defined as

$$\ddot{\eta}_i(\theta) = \frac{\partial^2 f(x_i; \theta)}{\partial\theta\partial\theta^T} = \left[\frac{\partial f^2(x_i; \theta)}{\partial\theta_j\partial\theta_k}\right]_{p\times p}, j = 1,\ldots,p; k = 1,\ldots,p \qquad (2.5)$$

We denote the second derivative of f at point x_i with respect to the jth and kth parameters θ_j and θ_k by f_{ijk}''. For computational purposes the Hessian matrix can be stored in a three-dimensional $(n \times p \times p)$ array, denoted by $\ddot{\eta}$,

$$\ddot{\eta} = [f_{ijk}'']$$

The tail-cross product defined below is a convergence of the product of sequences. It will be used in the consistency and asymptotic normality of estimators.

Definition 2.4 *Tail-cross products* Let the functions $g(\alpha)$ and $h(\beta)$ be two sequence valued functions defined on Θ. Let $\langle g(\alpha), h(\beta)\rangle_n = n^{-1}\sum_{i=1}^{n} g_i(\alpha)h_n(\beta)$. If $\langle g(\alpha), h(\beta)\rangle_n \rightarrow \langle g(\alpha), h(\beta)\rangle$ uniformly for all α and β on Θ, let $[g(\alpha), h(\beta)]$ denote the function on $\Theta \times \Theta$ that takes $\langle \alpha, \beta\rangle$ into $\langle g(\alpha), h(\beta)\rangle$. This function is called the tail-cross product of g and h.

2.3 Parameter Estimations

The parameter $\theta \in \mathfrak{R}^p$ of a nonlinear regression model is a vector of p unknown values, which have to be estimated. There are several ways to estimate the parameters. The estimators in this chapter are based on classical regularity assumptions about errors. Let the following condition hold:

$$\varepsilon \sim N(0, \sigma^2 I)$$

that is, the errors are independent with constant variance and normal distribution. The likelihood estimation can then be used for estimation and inference purposes. In this case the least squares estimator is equivalent to maximum likelihood estimates (MLEs).

2.3.1 Maximum Likelihood Estimators

If the errors ε_i are i.i.d. in a normal distribution with equal variance, we can estimate the parameter vector θ using MLEs. The likelihood function can be written as:

$$\varepsilon_i \sim N(0, \sigma^2)$$

$$\Rightarrow L(\theta, \sigma) = \prod_{i=1}^{n} \frac{1}{\sqrt{2\pi\sigma^2}} \exp\left\{ -\frac{1}{2}\left(\frac{y_i - f(x_i; \theta)}{\sigma}\right)^2 \right\} \tag{2.6}$$

and the parameter estimates of (θ, σ) can be computed by maximizing the likelihood function. It is equivalent and mathematically easier to maximize the log likelihood $\ell(\theta, \sigma) = \log(L)$, or minimize the negative of the log likelihood.

$$\ell(\theta, \sigma) = -\sum_{i=1}^{n}\left\{ \log(\sigma^2) + \left(\frac{y_i - f(x_i; \theta)}{\sigma}\right)^2 \right\}$$

$$= -n\log(\sigma^2) - \sum_{i=1}^{n}\left(\frac{y_i - f(x_i; \theta)}{\sigma}\right)^2 \tag{2.7}$$

$$(\hat{\theta}, \hat{\sigma})_{MLE} = \arg\max_{(\theta, \sigma)} \ell(\theta, \sigma) \tag{2.8}$$

where constant terms are dropped. The MLE is efficiently normally distributed, with mean zero and a covariance matrix that can be written in terms of the Fisher information:

$$(\hat{\theta}, \hat{\sigma})_{MLE} \sim N(0, \frac{1}{n}I^{-1}) \tag{2.9}$$

The Fisher information is given by

$$I = E\left(\frac{\partial}{\partial(\theta, \sigma)}\ell(\theta, \sigma)\right)^2$$

$$= -E\left(\frac{\partial^2}{\partial(\theta, \sigma)(\theta, \sigma)^T}\ell(\theta, \sigma)\right) \tag{2.10}$$

in which the derivatives of the log-likelihood function (2.7), for a nonlinear regression model, come in the form:

$$\frac{\partial}{\partial\theta}\ell(\theta, \sigma) = \frac{1}{\sigma^2}\sum_{i=1}^{n}\left(\frac{\partial}{\partial\theta}f(x_i; \theta)\frac{y_i - f(x_i; \theta)}{\sigma^2}\right)$$

$$= \frac{1}{\sigma^2}\dot{\eta}(\theta)^T \mathbf{r}(\theta)$$

The derivative with respect to variance (σ^2) is

$$\frac{\partial}{\partial\sigma^2}\ell(\theta, \sigma) = \frac{-n}{2\sigma^2} + \frac{1}{2(\sigma^2)^2}\sum(r_i^2(\theta))$$

and the second derivatives are

$$\frac{\partial^2}{\partial\theta\partial\theta^T}\ell(\theta, \sigma) = \frac{1}{\sigma^2}\ddot{\eta}(\theta)\mathbf{r}(\theta) - \frac{1}{\sigma^2}\dot{\eta}^T(\theta)\dot{\eta}(\theta)$$

$$\frac{\partial^2}{\partial(\sigma^2)^2}\ell(\theta, \sigma) = \frac{n}{2(\sigma^2)^2} - \frac{1}{(\sigma^2)^3}\mathbf{r}^T(\theta)\mathbf{r}(\theta)$$

$$\frac{\partial^2}{\partial\theta\partial\sigma}\ell(\theta, \sigma) = -\frac{1}{\sigma^4}\eta(\theta)^T\mathbf{r}(\theta)$$

Note that $E(\mathbf{r}(\theta)) = 0$, so we can conclude that the Fisher information matrix is

$$I = \begin{bmatrix} \frac{1}{\sigma^2} \dot{\eta}(\theta_0)^T \dot{\eta}(\theta_0) & 0 \\ 0 & \frac{n}{2\sigma^4} \end{bmatrix} \tag{2.11}$$

2.3.2 The Ordinary Least Squares Method

The ordinary least squares (OLS) method is widely used to estimate the parameters of a nonlinear model. Denoting the estimator by $\hat{\theta}_{OLS}$ minimizes the sum of squared errors

$$S(\theta) = \sum_{i=1}^{n} [y_i - f(x_i; \theta)]^2 \tag{2.12}$$

$$= [\mathbf{y} - \eta(\theta)]^T [\mathbf{y} - \eta(\theta)] \tag{2.13}$$

over parameter vector θ in Θ, which is a subset of the Euclidean space $\Theta \subset \mathfrak{R}^p$. The covariance matrix of the least squares estimate will be equivalent to the MLE if the error follows a normal distribution:

$$Cov(\hat{\theta}_{OLS}) = Cov(\hat{\theta}_{MLE}) = \sigma^2 [\dot{\eta}(\theta_0)^T \dot{\eta}(\theta_0)]^{-1} \tag{2.14}$$

where $\dot{\eta}$ is gradient vector of a nonlinear model function with respect to the parameter defined by (2.3).

The least squares estimate of the error variance σ^2 is

$$\hat{\sigma} = \frac{S(\hat{\theta})}{n - p}$$

Note that the parameter covariance matrix is unknown and can be estimated by replacing the true parameter θ_0 by its estimate (for simplicity we denote $\hat{\theta}_{OLS}$ as $\hat{\theta}$):

$$\widehat{Cov}(\hat{\theta}_{OLS}) = \hat{\sigma}^2 [\dot{\eta}(\hat{\theta})^T \dot{\eta}(\hat{\theta})]^{-1} \tag{2.15}$$

Under certain conditions the existence, consistency, and asymptotic normality of least squares estimators can be proved. First, assume the following:

A 2.5 Assume the model (2.1) is satisfied. Let $f(x_i; \theta)$ be known continuous functions on a compact subset Θ of a Euclidean space, and errors ε_i be i.i.d. with zero mean and finite variance $\sigma^2 > 0$.

The following theorem, proved by Jennrich (1969), shows that A2.5 is a sufficient condition to guarantee the existence of the least squares estimator.

Theorem 2.6 Let $S(\theta)$, the sum of squared errors, be a real-valued function on $\Theta \times \mathbf{Y}$, where $\mathbf{Y}$ is a measurable space. For each $\theta \in \Theta$, let $S(\theta, y)$ be a measurable function of y and each $y \in \mathbf{Y}$ be a continuous function of θ. Then there exists a measurable function $\hat{\theta}$ from $\mathbf{Y}$ into Θ that for all $y \in \mathbf{Y}$,

$$S(\hat{\theta}(y), y) = \inf_{\theta} S(\theta, y)$$

The importance of the OLS estimator is not only its ease of use but also because under certain conditions it has useful mathematical properties, such as strong consistency, strong convergence, and asymptotic efficiency. Moreover, if the errors are normally distributed, least squares estimates become MLEs. Furthermore, the Gauss–Newton iteration method for numeric computation of the OLS is asymptotically stable. There are several regularity conditions that are required for strong consistency.

A 2.7   Assume that the limit of the sequence $\frac{1}{n}\sum_{i=1}^{n}f^2(x_i; \theta)$ (tail-cross product of f_i with itself) exists, and that

$$Q(\theta) = |f(\theta) - f(\theta_0)|^2$$

has a unique minimum at a true value θ_0 of an unknown parameter θ.

Theorem 2.8   Let $\hat{\theta}_n$ be a sequence of least squares estimators. Under A 2.5 and A 2.7, $\hat{\theta}_n$ and $\hat{\sigma}_n^2 = \frac{1}{n}S(\hat{\theta})$ are strongly consistent estimators of θ_0 and σ^2.

For the proof see Jennrich (1969). The asymptotic normality of OLS requires some more assumptions:

A 2.9   Assume the derivatives f_{ij}' and f_{ijk}'' exist and are continuous on Θ, and that all tail-cross products of the form $\langle g, h \rangle$, where $g, h = f, f', f''$, exist. These are defined as:

$$a_{nij}(\theta) = <f_i'(\theta), f_j'(\theta)>_n \quad A_n(\theta) = [a_{nij}\theta]$$

$$a_{ij}(\theta) = <f_i'(\theta), f_j'(\theta)> \quad A(\theta) = [a_{ij}\theta]$$

A 2.10   The true vector θ_0 is the interior point of Θ, and the matrix of the derivative tail-cross product

$$A_n(\theta) = \frac{1}{n}\dot{\eta}^T\dot{\eta} \to A(\theta)$$

is nonsingular.

Under A2.5–A2.10, Jennrich (1969) proved the asymptotic normality of OLS.

Theorem 2.11   Let $\hat{\theta}_n$ be the sequence of least squares estimators of θ_0. Under A2.5–A2.10 (Jennrich 1969)

$$D(n^{\frac{1}{2}}(\hat{\theta}_n - \theta_0)) \to N(0, \sigma^2 A^{-1}) \tag{2.16}$$

Moreover, $A_n(\hat{\theta})$ is a strong estimator of $A(\theta_0)$.

2.3.3 Generalized Least Squares Estimate

The weighted or generalized least squares (GLS) estimate is the answer to the minimization of the function given by

$$S(\theta) = [\mathbf{Y} - \boldsymbol{\eta}]^T V [\mathbf{Y} - \boldsymbol{\eta}] \tag{2.17}$$

where V is a known positive definite matrix. In the nonlinear model 2.2, let the expectation of error be zero ($E(\boldsymbol{\varepsilon}) = \mathbf{0}$), and the covariance matrix of errors be a known matrix given by $Cov(\boldsymbol{\varepsilon}) = \sigma^2 V$. The OLS estimator discussed above is a special case in which $V = I_n$. Let $V = U^T U$ be the Cholesky decomposition of V, where U is an upper triangular matrix. Multiplying the model through by $R = (U^T)^{-1}$ transforms it to

$$\mathbf{Z} = \boldsymbol{\kappa} + \boldsymbol{\xi} \tag{2.18}$$

where $\mathbf{Z} = R\mathbf{y}$, $\boldsymbol{\kappa} = R\boldsymbol{\eta}$, and $\boldsymbol{\xi} = R\boldsymbol{\varepsilon}$. Then $E(\boldsymbol{\xi}) = \mathbf{0}$ and $Cov(\boldsymbol{\xi}) = \sigma^2 RVR^T = \sigma^2 I_n$. Thus the GLS is transformed into an OLS model, and the sum of squares is

$$S(\theta) = [\mathbf{y} - \boldsymbol{\eta}]^T R^T R [\mathbf{y} - \boldsymbol{\eta}] \tag{2.19}$$

$$= [\mathbf{Z} - \boldsymbol{\kappa}]^T [\mathbf{Z} - \boldsymbol{\kappa}] \tag{2.20}$$

that is, the ordinary sum of squared errors, hence GLS is the same as OLS for the transformed model. Consequently, all the OLS formulas can be derived from the transformed model.

The gradient matrix $\dot{\boldsymbol{\kappa}} = \partial \boldsymbol{\kappa} / \partial \theta$ can be written as:

$$\dot{\boldsymbol{\kappa}} = R\dot{\boldsymbol{\eta}},$$

so the covariance matrix of parameters is equal to

$$Cov(\hat{\theta}_{GLS}) = \sigma^2 [\dot{\boldsymbol{\kappa}}(\theta_0)^T \dot{\boldsymbol{\kappa}}(\theta_0)]^{-1}$$

$$= \sigma^2 [\dot{\boldsymbol{\eta}}^T(\theta_0) V^{-1} \dot{\boldsymbol{\eta}}(\theta_0)]^{-1}$$

This matrix can be estimated by

$$\widehat{Cov}(\hat{\theta}_{GLS}) = \hat{\sigma} [\dot{\boldsymbol{\eta}}(\hat{\theta}_{GLS}) V^{-1} \dot{\boldsymbol{\eta}}(\hat{\theta}_{GLS})] \tag{2.21}$$

where

$$\hat{\sigma}^2 = \frac{1}{n-p} [\mathbf{y} - \boldsymbol{\eta}(\hat{\theta}_{GLS})]^T V^{-1} [\mathbf{y} - \boldsymbol{\eta}(\hat{\theta}_{GLS})]$$

In this book, this generalized form is used when the classical assumptions, such as independence and homogeneity of error variances, do not hold. In this case, the covariance matrix part (V) includes the heterogeneous variance on the diagonal (see Chapter 4); the off-diagonal values are the autocovariances of the errors (see Chapter 5). In such cases, the covariance matrix might depend on unknown parameters, which we can attempt to estimate. In this case the procedure is first to estimate the parameter vector using OLS, then to estimate the covariance matrix, and finally to estimate the parameters using GLS.

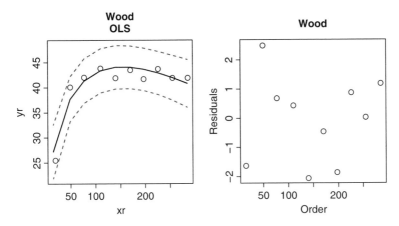

Figure 2.1 Fitted Wood model for a single cow's milk production data for a year.

2.4 A Nonlinear Model Example

Example 2.1 The cow milk data shown in Table A.8 (Appendix A) were obtained from a dairy farm. The full data set is recorded for each of 177 cows over a year of producing milk. The full data follow mixed models, but in this example only the data for a single cow are presented to find a nonlinear model. Silvestre et al. (2006) applied seven mathematical functions for modeling such data. The Wood model is fitted to these data and is defined as:

$$y_i = ax_i^b e^{-cx_i} + \varepsilon_i; \ i = 1, \ldots, n \tag{2.22}$$

In the model, parameter a is a scaling factor to represent yield at the beginning of lactation, and parameters b and c are factors associated with the inclining and declining slopes of the lactation curve, respectively. The OLS estimate of the parameter vector is equal to ($\hat{a} = 11.262, \hat{b} = 0.338, \hat{c} = 2.185e - 3$), with error variance $\hat{\sigma}^2 = 1.675$. Figure 2.1 shows the fitted curve and residual plot.

The correlation matrix, computed using Equation 2.15, is equal to:

$$Cov[\hat{a}, \hat{b}, \hat{c}]^T = \begin{bmatrix} 1.000 & -0.986 & -0.856 \\ & 1.000 & 0.926 \\ & & 1.000 \end{bmatrix}$$

The R-tools for fit and parameter inference are presented in Chapter 8.

3

Robust Estimators in Nonlinear Regression

Classical assumptions – independence, homogeneity of error variances, normality of errors, and many others – might not be satisfied in real-life data. Some of these null conditions, such as heteroscedasticity of error variances and autocorrelated errors, will be discussed in later chapters.

The classical assumptions of a regression model may not be met in real-life data. In particular, outliers may breach those classical assumptions; their effect will therefore be to mislead. To show these effects, Figure 3.1 displays some simulated data generated from the logistic model $y = a/(1 + b\exp(-cx)) + \varepsilon$, in which the errors ε are generated from an independent normal distribution with constant variance. Several artificial outliers are then added to the data and parameters are estimated using the classical least squares estimator.

The plot not only shows that the parameter estimates are wrong, but also mistakenly suggests the errors are dependent, with nonconstant variance. The outliers shift the curve towards them, and the residual plot wrongly displays a systematic pattern.

This shows that robust methods suitable for fitting the models are required not only when the classical assumptions are satisfied, but also in cases of error variance heterogeneity and autocorrelated errors. Moreover, suitable outlier detection methods are also required for such situations. In this chapter, robust estimating methods are discussed for when the classical assumptions of the regression model holds. In Chapter 4 the heteroscedasticity case is explained, in Chapter 5 the autocorrelated error case is studied, and in Chapter 6 methods for outlier detection in nonlinear regression are presented.

To complete this example, Figure 3.1 also shows the robust fit of the simulated example, in which the above problems are resolved.

3.1 Outliers in Nonlinear Regression

The simulated example data in Table A.10, and shown in Figure 3.1, demonstrate that outliers can have more unpredictable effects in nonlinear than in linear regressions. For example, Figure 3.2 uses the same simulated data structure

Robust Nonlinear Regression: with Applications using R, First Edition.
Hossein Riazoshams, Habshah Midi, and Gebrenegus Ghilagaber.

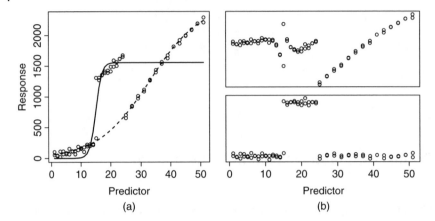

Figure 3.1 Simulated logistic model, artificially contaminated. (a) Logistic model: solid line, least squares estimate; dashed line, robust MM-estimate. (b) Computed residuals. Upper chart: least squares estimate; lower chart: MM-estimate.

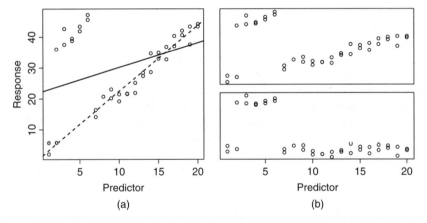

Figure 3.2 Simulated linear model, artificially contaminated. (a) Least squares fit: solid line: least squares estimate; dashed line: robust MM-estimate. (b) Computed residuals: top, least squares estimate; bottom, MM-estimate.

(y direction) in a linear regression. The position of the outliers can have different effects on a specific parameter in both linear and nonlinear regressions. The difference is that a linear regression does not bend in the presence of outliers, but rotates and moves due to changes in the intercept and slope. The residual plots computed by least squares in Figure 3.2 therefore display a few suspicious points that might be outliers, but this is not be true in all cases. However, many researchers have incorrectly used the residual plot from a classical fit to

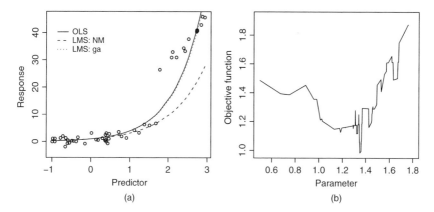

Figure 3.3 Simulated nonlinear exponential model, artificially contaminated. (a) Exponential model: solid line, estimates using OLS; dotted line, LMS using Nelder–Mead optimization; dashed line, LMS using a genetic algorithm. (b) Median of error squares: objective function LMS for different parameter values.

identify outliers. This may sometimes (as in this example) help, but in nonlinear regression, as in Figure 3.3, the curve bends towards the outlier points. This gives misleading results, not only in outlier detection but also in parameter estimation.

3.2 Breakdown Point in Nonlinear Regression

Stromberg and Ruppert (1992) extended the finite breakdown point concept 1.8 to nonlinear regression. They defined breakdown in terms of the estimated regression function, not the estimated parameter.

Let $\mathbb{X}$ be the space of predictor $x \in \mathfrak{R}^k$ values. For any $x \in \mathbb{X}$ and estimator $\hat{\theta}$, the values of the upper and lower breakdown functions are defined as the lowest proportions of contamination that can give the estimated regression function to $\sup_\theta f(x; \theta)$ and $\inf_\theta f(x; \theta)$, respectively. More specifically, at $x \in \mathbb{X}$, for regression function, sample parameter estimate $\hat{\theta}_n$, and sample $X_n = (x_1, \ldots, x_n)$, the upper breakdown function is defined as:

$$\varepsilon_+(x, f, \hat{\theta}, X_n)$$

$$= \min_{0 \leqslant m \leqslant n} \left\{ \frac{m}{n} \sup_{x \in \mathfrak{R}^k} f(x; \hat{\theta}_n) = \sup_\theta f(x; \theta) \right\}$$

$$\text{if } \sup_\theta f(x; \theta) > f(x; \hat{\theta}_n)$$

$$= 1 \quad \text{otherwise}$$

The lower breakdown function is defined as:

$$\varepsilon_-(x,f,\hat{\theta},X_n)$$

$$= \min_{0 \leqslant m \leqslant n} \left\{ \frac{m}{n} \inf_{x \in \mathfrak{R}^{k^*}} f(x; \hat{\theta}_n) = \inf_{\theta} f(x; \theta) \right\}$$

$$\text{if} \quad \inf_{\theta} f(x; \theta) > f(x; \hat{\theta}_n)$$

$$= 1 \quad \text{otherwise}$$

The breakdown function is defined as:

$$\varepsilon(x,f,\hat{\theta},X_n) = \min\{\varepsilon_+(x,f,\hat{\theta},X_n), \varepsilon_-(x,f,\hat{\theta},X_n)\}$$

Finally, the finite breakdown point is defined as:

$$\varepsilon(f,\hat{\theta},X_n) = \inf_{x \in \mathbb{X}} \{\varepsilon(x,f,\hat{\theta},X_n)\}$$

3.3 Parameter Estimation

In this chapter we assume the nonlinear regression model

$$y_i = f(x_i; \theta) + \varepsilon_i, \quad i = 1, \ldots, n$$

holds, and errors are independent with constant variance. The parameter $\theta \in \mathfrak{R}$ and variance of errors $(Var(\varepsilon_i) = \sigma^2)$ are unknown and have to be estimated. Several robust methods of estimation are discussed in this chapter.

3.4 Least Absolute and Quantile Estimates

The first method of finding a robust estimator comes from Edgeworth (1887), who proposed replacing the squares of errors by absolute values. Analogous to linear regression (Section 1.9), the least absolute (LA) value estimates minimize the l_1 criterion, defined as

$$\hat{\theta}_{LA} = \arg\min_{\theta} \sum_{i=1}^{n} |r_i(\theta)| \tag{3.1}$$

for defined errors

$$r_i(\theta) = y_i - f(x_i; \theta), i = 1, \ldots, n \tag{3.2}$$

This protects the estimate against y_i, but still cannot cope with outlying x_i, which are referred to as "leverage points".

The minimization problem can be reformulated as a linear programming problem via a linear approximation of the nonlinear regression. As noted by

Attar et al. (1979), a necessary condition for $\hat{\theta}$ to solve (3.1) is that there exists a vector $\mathbf{d} \in [-1, 1]^n$ such that

$$\dot{\eta}(\hat{\theta})^T \mathbf{d} = 0$$

$$\eta(\hat{\theta})^T \mathbf{d} = \sum |\eta(\hat{\theta})|$$

Thus, as proposed by Osborne and Watson (1971), one approach to solving (3.1) is to solve a succession of linearized l_1 problems minimizing

$$\sum (\eta_i(\theta) - \dot{\eta}^T \delta)$$

and choosing a step length λ, at each iteration, by line search direction δ. Detailed discussion of this approach is given by Koenker and Park (1996).

3.5 Quantile Regression

As in the case of the linear problem, the generalization of the l_1 problem to other quantiles is straightforward, involving only a modification of the constraint set $[-1, 1]^n$ to $[\alpha - 1, \alpha]$ for some $\alpha \in (0, 1)$, where α appears only in the recentering of the $\mathbf{d}$ vector.

The linear criterion gives the estimate of θ as

$$\hat{\theta}_\alpha = \arg\min_\beta = \sum_{i=1}^{n} \rho_\alpha[y_i - f(X_i; \theta)] \, \rho_\alpha(u) = y(\alpha - I(u < 0)) \tag{3.3}$$

The quantile regression of Koenker and Park (1996) is obtained by the dual problem:

$$\max\{Y'\mathbf{d} | \mathbf{d} \in \Omega = d \in [\alpha - 1, \alpha]^n, X'\mathbf{d} = 0\}.$$

Quantile regression examples in R are presented in Chapter 9.

3.6 Least Median of Squares

The least median of squares (LMS) estimator was defined by Rousseeuw (1984) for linear regression. It provides the answer to the problem of minimizing the median of squared errors. It is HBP, but not necessarily efficient.

The LMS estimate is the set of parameter values that minimize the median of squared errors loss function given by

$$\ell_{LMS}(\theta) = \underset{1 \le i \le n}{\text{Median}}(r_i^2(\theta)) \tag{3.4}$$

and provides the answer to the minimization problem:

$$\hat{\theta}_{LMS} = \arg\min_{\theta \in \Theta} \ell_{LMS}(\theta) \tag{3.5}$$

The LMS estimate is HBP, but is not unique and efficient. Stromberg (1995) proved the weak convergence of LMS. Since it is HBP, it is a suitable initial value for the S-estimate or MM-estimate, as discussed later in this chapter.

The notation is as follows:

- $r_i = y_i - f(x_i; \theta), i = 1, \ldots, n \equiv$ errors
- $\mathbf{r}^T(\theta) = [r_1, \ldots, r_n] \equiv$ vector of errors
- $F_{\mathbf{r}^2(\theta)} \equiv$ distribution function of squared errors $\mathbf{r}^T(\theta)$
- $Q_q \equiv q$th quantile of the distribution of squared errors $F_{\mathbf{r}^2(\theta)}$

Let the following assumptions hold:

A 3.1 Define $m(\theta) = \text{Median}(\mathbf{r})$ and assume $\forall \varepsilon > 0$, $m(\theta_0) < \inf_{\|\theta - \theta_0\| > \varepsilon} m(\theta)$.

A 3.2 $\forall x$ in sample space of X_i and $\forall \theta \in \Theta, f(x, ; \theta)$ is jointly continuous in x and θ.

A 3.3 $\forall \theta \in \Theta$, $\forall q \in (0, 1] : Q_q(\theta)$ is the unique solution of

$$F_{R^2(\theta)}(Q_q(\theta)^-) \le p \le F_{R^2(\theta)}(Q_q(\theta))$$

Stromberg (1995) proved that under A3.1–A3.3 the LMS is weakly convergent. The LMS estimator is HBP (50%), but it is not unique. For example, Figure 3.3 displays a simulated regression from the exponential model $y = e^{ax} + \varepsilon \sim N(0, 1)$ with a single parameter $a = 1.1$. The LMS and MM-estimate remained robust even after contamination of 28% of the data. The objective function of the median error squares given by (3.4) is shown in Figure 3.3. It shows the least median error square object function is not smooth, including several local minima, and therefore gives nonunique estimates.

The HBP property, but not necessarily the efficiency, of LMS makes it a suitable starting value for other efficient estimators, such as the MM- and S-estimators, which require an HBP estimate as their initial values. Moreover, in nonlinear regression, most estimators have to be computed by numerical methods, which require initial values their first iterations. Sensitivity to initial values and outliers affects the results of these numerical computations. Unsuitable initial values and the presence of outliers can cause break down even of an efficient robust estimator. Thus, a good starting value in a nonlinear regression is critical, especially when we do not have any idea what likely parameter values are.

In addition to the parameter vector θ, the scale parameter σ has to be estimated in a robust way. Rousseeuw (1984) proposed estimating the scale for linear regression by

$$\hat{\sigma}_{LMS} = 1.4826 c(n, p) \ell(\hat{\theta}_{LMS})$$

where $1.4826 \simeq 1/\Phi^{-1}(0.75)$ is an asymptotic correction factor for the case of normal errors. The constant $c(n,p)$, which is a finite sample correction factor larger than 1, is used to make the estimate unbiased when simulating samples with normal errors. Vankeerberghen et al. (1995) discussed the LMS estimate in nonlinear regression, and applied a genetic algorithm for computational purposes. They obtained the pure scale σ robust estimate in two steps. In the first step, an initial scale estimate is calculated by means of the equation

$$S = 1.4826 \left(1 + \frac{5}{n-p} \right) \sqrt{\mathrm{Median}[r_i^2(\hat{\theta}_{LMS})]}$$

where $r_i^2(\hat{\theta}_{LMS})$ is the LMS residuals. In the second step, the LMS residuals of the data points are compared with this initial scale estimate. When a residual is larger than 2.5 times S, the data point is omitted. If n^* is the number of points retained, the final scale estimate is calculated using the equation

$$\hat{\sigma}^2 = \sqrt{\frac{\sum_{i=1}^{n^*} r_i^2(\hat{\theta}_{LMS})}{n^* - p}} \tag{3.6}$$

This method is used in this book and the `nlr` package is used to compute the LMS robust estimate of scale, as discussed in Chapter 8.

3.7 Least Trimmed Squares

The least trimmed squares (LTS) estimate, denoted by $\hat{\theta}_{LTS}$, was proposed by Rousseeuw (1984,1985), who studied its robustness and consistency for linear regressions. Stromberg and Ruppert (1992) discussed the robust properties of LTS in nonlinear regression. Chen et al. (1997) and Čížek (2001) proved the consistency and asymptotic normality of the LTS estimate in nonlinear regression.

The approach is based on eliminating points with high residual values, then computing the least squares estimate for the rest of the data. The LTS loss function is defined as:

$$\ell_{LTS}(\theta) = \sum_{i=1}^{h} r_{i:n}^2(\theta), \quad n/2 \leqslant h \leqslant n$$

and the LTS estimate (denoted by $\hat{\theta}$) is the result of the minimization problem:

$$\hat{\theta} = \arg\min_{\theta \in \Theta} \ell_{LTS}(\theta)$$

where $r_{1:n}^2 \leqslant \ldots, \leqslant r_{n:n}^2$ are the ordered squared residuals. The sum is taken over h smallest terms, where h determines the robustness and efficiency of the LTS estimator.

Let $\alpha = 1 - 1/n$ be the amount of trimming (the "trimming proportion"), with $0 \leqslant \alpha \leqslant 1/2$. The maximum breakdown point (BP) for LTS equals $\frac{1}{2}$ and is obtained by choosing α close to $\frac{1}{2}$. However, one may expect a trade-off between a high value of α and a loss in efficiency. Therefore, the choice of α (or equivalently h) determines the overall performance of the LTS estimator, and some effort in tuning this parameter is therefore required. Hence, it has been suggested that lower values for α will give a good compromise between robustness and efficiency. The most commonly suggested values are 0.25 and 1.

LTS appears impractical for nonlinear regression. All of the estimation methods in nonlinear regression need parameter starting values, and these can be far from the actual values. The first few iterations will therefore face masking and swamping effects and whole iteration procedure will break down.

The solution is to start from an LMS estimate. However, by removing some of the data in a nonlinear regression, the model will be fitted to a smaller range of data, which is an incorrect modeling approach: generally in nonlinear regression, the shape of the curve depends on the position of the points, so removing data might cause the estimator to end up with an incorrect model fit. There are several examples of this in nonlinear regression and in parameter estimates in other cases.

3.8 Least Trimmed Differences

The least trimmed differences (LTD) method was introduced by Stromberg et al. (2000) for linear regression and they studied its asymptotic normality properties. The estimator minimizes the sum of the smallest quartile of the squares of the differences in each pair of residuals. The LTD loss function is defined as:

$$\ell_{LTD}(\theta) = \sum_{k=1}^{\binom{h}{2}} \{[r_i(\theta) - r_j(\theta)]^2; i < j\}_{k:\binom{n}{2}}$$

where the pairwise differences extend over all values of i and j, from 1 to n, for which $i < j$, and the trimmed constant h has to satisfy $n/2 \leqslant h \leqslant n$. The notation $k : \binom{n}{2}$ means kth-order statistics among the $\binom{n}{2}$ elements of the set $\{[r_i(\theta) - r_j(\theta)]^2; i < j\}$.

Note that the objective function of LTD is discontinuous, nonconvex, and not differentiable, the same as for both LMS and LTS. Therefore these optimization problems cannot be solved by standard derivative-based methods.

3.9 S-estimator

The S-estimate was proposed by Rousseeuw and Yohai (1984) for linear regression. They discussed its breakdown property, consistency, and asymptotic normality (see Section 1.8). Sakata and White (2001) investigated the large-sample properties of S-estimators in nonlinear regressions with dependent and heterogeneous data in a time series.

The S-estimate is based on the M-estimate for the scale parameter σ. It is used in next section to define the τ-estimate.

The definition and discussion of linear regression in Section 1.8 can be extended to nonlinear regression. Define the dispersion function $S(r_1, \ldots, r_n)$ to be a function of any sample $(r_1, \ldots, r_n)$ of real numbers, as in the solution of the Huber equation of the form:

$$\sum_{i=1}^{n} \rho\left(\frac{r_i(\theta)}{S}\right) = nE_\phi(\rho) \tag{3.7}$$

where $E_\phi(\rho)$ is the mean of the real-valued function ρ and ϕ is the standard normal density function. The real-valued function ρ was defined in Chapter 1. Let A.1.27–A.1.30 hold. To estimate the function model parameter θ and scale σ for each vector θ, we obtain residuals $r_i(\theta) = y_i - f(x_i; \theta)$, then calculate the dispersion function $S(r_1(\theta), \ldots, r_n(\theta))$ using Equation 3.7. After that, we define the S-estimate of θ by:

$$\hat{\theta}_{n,S} = \arg\min_\theta S(r_1(\theta), \ldots, r_n(\theta)) \tag{3.8}$$

The final scale estimator is

$$\hat{\sigma}_{n,S} = S(r_1(\hat{\theta}_S), \ldots, r_n(\hat{\theta}_S))$$

S-estimators are affine equivariant. If $\frac{E_\phi(\rho)}{\rho(c)} = \lambda$ and A.1.27–A.1.28 hold, the breakdown of the S-estimate tends to λ.

Although the computation of the S-estimate is difficult, because the scale estimate is in fact the M-estimate, we can derive the constants of robust loss ρ functions. Recall that to achieve the HBP estimate A.1.30 must hold. This can be used to compute the consistency factor constant c. For example, consider the Tukey bi-square function in Table A.11:

$$\rho(t) = \begin{cases} 1 - \left(1 - \left(\frac{t}{\alpha}\right)^2\right)^3 & |t| \leq \alpha \\ 1 & \alpha < |t| \end{cases} \tag{3.9}$$

We have to calculate α such that $\frac{E_\phi(\rho)}{\rho(\alpha)} = \frac{1}{2}$. Note that (3.9) can be simplified as

$$\rho(t) = \left(3\frac{t^2}{a^2} - 3\frac{t^4}{a^4} + \frac{t^6}{a^6} \right) I[|t| < \alpha] + I[|t| > \alpha]$$

where I stands for an indicator function and $\rho(\alpha) = 1$. Therefore, the following expression must hold:

$$\int_{-\alpha}^{\alpha} \left(3\frac{t^2}{a^2} - 3\frac{t^4}{a^4} + \frac{t^6}{a^6} \right) d\Phi(t) + P_\Phi(|t| > \alpha) = \lambda \tag{3.10}$$

where $\lambda = 0.5$ guarantees the S-estimate to be HBP and includes $\alpha = 1.54765$. Several discussions around computation of (3.10) can be found in the literature, for example see Riani et al. (2014).

3.10 τ-estimator

Yohai and Zamar (1988) proposed a high-breakdown, high-efficiency estimator called the τ-estimate. Tabatabai and Argyros (1993) extended this to nonlinear regression, and discussed its inferences, asymptotic normality, and computing algorithm. The robust scale of the residuals is defined as:

$$\tau(\boldsymbol{\theta}, s_n) = s_n \sqrt{\frac{1}{n} \sum_{i=1}^{n} \rho_1 \left(\frac{r_i(\boldsymbol{\theta})}{s_n} \right)} \tag{3.11}$$

τ-estimates are defined by minimizing the robust scale of the residuals:

$$\hat{\boldsymbol{\theta}}_\tau = \arg\min_{\boldsymbol{\theta}} \tau(\boldsymbol{\theta}, s_n) \tag{3.12}$$

subject to the constraint

$$\frac{1}{n} \sum_{i=1}^{n} \rho_0 \left(\frac{r_i(\boldsymbol{\theta})}{s_n} \right) = b \tag{3.13}$$

s_n in Equation 3.13 is in fact the S-estimate of scale, with estimating Equation 3.7. The tuning constant b satisfies $b/\rho_0(\infty) = 0.5$, and together with the choice of ρ_0 regulates the robustness. The choice of ρ_1 can be tuned to give good asymptotic efficiency under the Gaussian model.

3.11 MM-estimate

The MM-estimate proposed by Yohai (1987) for linear regression was shown in Algorithm 1.4. The author proved that it is HBP and has asymptotic normality (see Section 1.7 for detailed discussion). Stromberg and Ruppert (1992) and Stromberg (1993) extended the MM-estimate to nonlinear regression. We can use Algorithm 1.4 with only slight changes to compute the MM-estimate for a nonlinear regression. In addition, we will use numerical computation methods that will be discussed in Chapter 7. Let A.1.13–A.1.15 of Algorithm 1.4 be satisfied, then the MM-estimate for a nonlinear regression model can be computed using the following algorithm:

Algorithm 3.1 Robust M-estimate for nonlinear regression.

Stage 1: Take an HBP and consistent estimate $\hat{\theta}_{0,n}$ of θ.

Stage 2: Compute the residuals $r_i(\hat{\theta}_{0,n}) = y_i - f(x_i; \theta_{0,n})$.

Stage 3: Compute the M-estimate of scale $\hat{\sigma}_n = \sigma(r_i(\hat{\theta}_{0,n}))$ using a function ρ_0 obtained from

$$\frac{1}{n} \sum_{i=1}^{n} \rho_0 \left(\frac{r_i(\hat{\theta}_{0,n})}{\hat{\sigma}_n} \right) = b, \tag{3.14}$$

b is constant such that

$$b = E_\phi(\rho_0(t)).$$

Let $\psi_1 = \rho'_1$. Then the MM-estimate $\hat{\theta}_{1,n}$ is defined as the solution of the optimization problem:

$$\hat{\theta}_{1,n} = \arg \min_{\theta} S_n(\theta) \tag{3.15}$$

where

$$S_n(\theta) = \sum_{i=1}^{n} \rho_1 \left(\frac{r_i(\theta)}{\hat{\sigma}_n} \right)$$

and $\rho_1(0/0)$ is defined as 0. Equivalently the parameter is the solution of derivative equation:

$$\sum_{i=1}^{n} \psi_1 \left(\frac{r_i(\theta)}{\hat{\sigma}_n} \right) \frac{\partial r_i(\theta)}{\partial \theta} = 0 \tag{3.16}$$

which verifies

$$S_n(\hat{\theta}_{1,n}) < S_n(\hat{\theta}_{0,n}) \tag{3.17}$$

Similar to Section 1.7, the constants a and b, and the robust loss functions ρ_0 and ρ_1, must satisfy A.1.13–A.1.15; that is:

A 3.4 $a = \max \rho_0(t)$ and $b/a = 0.5$

A 3.5 $\rho_1(t) \leq \rho_0(t)$

A 3.6 $\sup \rho_1(t) = \sup \rho_0(t) = a$

3.4 implies that the scale estimate $\hat{\sigma}$ has HBP of 50%. The distribution of parameters can be extended from linear to nonlinear regression using Equation 1.46. Let $\mathbf{z}_i = (y_i, X_i)$, $i = 1, \ldots, n$ be i.i.d. observations of a nonlinear regression. Let $G_0(X)$ be the distribution of X_i and F_0 be the distribution of error $\varepsilon_i = y_i - f(X_i; \theta)$. The distribution of $\mathbf{z}_i$ is then given by:

$$H_0(\mathbf{z}) = G_0(\mathbf{X})F_0(y_i - f(X_i; \theta_0))$$

where θ_0 is the true value of θ. Similarly, denote the true value of σ by σ_0. The limit distribution of the MM-estimate can be written as

$$n^{1/2}(\hat{\theta}_n - \theta_0) \to_d N(0, \sigma_0^2 [A(\psi_1, F_0)/B^2(\psi_1, F_0)]V^{-1}) \tag{3.18}$$

where V can be calculated using the gradient matrix, $V = \dot{\eta}^T(\theta)\dot{\eta}(\theta)$, which can be estimated by $V = \dot{\eta}^T(\hat{\theta}_n)\dot{\eta}(\hat{\theta}_n)$. In addition, we have

$$A(\psi, F) = E_F(\psi^2(u/\sigma_0))$$

and

$$B(\psi, F) = E_F(\psi'(u/\sigma_0))$$

These values are unknown and can be estimated by replacement with the estimated parameters. Therefore, the covariance matrix of the MM-estimator can be computed as:

$$\widehat{Cov}(\hat{\theta}_{MM}) = \hat{\sigma}_n \frac{[1/(n-p)] \sum \psi_i^2}{[(1/n) \sum \psi_i]^2} (\dot{\eta}(\hat{\theta})^T \dot{\eta}(\hat{\theta}))^{-1} \tag{3.19}$$

The possible choices of ρ_0 and ρ_1 to satisfy the regularity conditions 3.5 and 3.6 are to select a single rho function with certain tuning constants. Let $0 < k_0 < k_1$, $\rho_0(t) = \rho(t/k_0)$, and $\rho_1(t) = \rho(t/k_1)$. The value of k_1 will determine the asymptotic efficiency of the estimate, and k_0 determines the HBP of the scale estimate.

Table A.11 presents several robust loss ρ functions and the appropriate tuning constants. These functions are provided by the `nlr` package (see Chapter 8). As an example of how to compute tuning constants, consider the simplified form of bi-square function in (3.9) used by Yohai (1987):

$$\rho(t) = \begin{cases} 1 - (1 - t^2)^3 & |t| \leq 1 \\ 1 & \alpha < |t| \end{cases} \tag{3.20}$$

Note that the bi-square function (3.20) at t/k_0 is equivalent to (3.9), with constant $c = k_0$. The consistency factor for the S-estimate evaluated in Section 3.10 was $c = 1.54765$. The maximum value of the bi-square function is $a = 1$, so to achieve the HBP in Assumption 3.4 implies $b = 0.5$. Thereafter, constant k_0 can be computed from $b = E_{\phi}\rho(t/k_0)$. Following the same argument as (3.10), we again conclude that $k_0 = 1.54765$.

The value of k_1 gives the efficiency and is independent of k_0. Consider the variance of the MM-estimate in (3.18). The efficiency of $\hat{\theta}_{MM}$, using the Fisher information in (2.11), can be computed as

$$
\begin{aligned}
eff(\theta_{MM}) &= \frac{1/I(\beta)}{Var(\hat{\beta}_{1,n})} \\
&= \frac{\sigma_0^2(\dot{\eta}^T\dot{\eta})^{-1}}{\sigma_0^2[A(\psi_1,F_0)/B^2(\psi_1,F_0)]V^{-1}} \\
&= \frac{B^2(\psi_1,F_0)}{A(\psi_1,F_0)}
\end{aligned}
$$

which depends on k_1 only. To achieve 95% efficiency, we have to calculate k_1 such that $eff(\theta_{MM}) = 0.95$. Without losing generality, we assume that $\sigma_0 = 1$, then

$$
A(\psi,F) = \int_{-k_1}^{k_1} \left[6\frac{t}{k_1} - 12\left(\frac{t}{k_1}\right)^3 + 6\left(\frac{t}{k_1}\right)^5\right]^2 d\Phi(t)
$$

and

$$
B(\psi,F) = \int_{-k_1}^{k_1} \left[6 - 36\left(\frac{t}{k_1}\right)^2 + 30\left(\frac{t}{k1}\right)^4\right] d\Phi(t)
$$

so by solving the equation $eff(\theta_{MM}) = 0.95$, we conclude that $k_1 = 1.677$ (1.676661 to be exact).

The MM-estimator is the basis of the estimators used in most of the methods covered in this book that will be developed in subsequent chapters to handle heteroscedasticity and autocorrelated errors. Furthermore, it is the basis of the `nlr` R package developed by the authors of this book.

When the errors do not comply with the independence and homogeneity assumptions, we can extend the generalized least squares approach discussed in Section 2.3.3 to the MM-estimate. Similarly, assuming that the covariance matrix of errors is a known value $Cov(\varepsilon) = \sigma^2 V$, using the Cholesky decomposition of $V = U^T U$ and the transposed inverse of $R = (U^T)^{-1}$, we can transfer the nonlinear regression model to (2.18) with constant and independent variance. Therefore, the generalized M-estimate (for known σ) can be defined as

$$
\hat{\theta}_M = \arg\min_{\theta} \sum_{i=1}^{n} \rho\left(\frac{Ry_i - Rf(x_i;\theta)}{\sigma}\right) \tag{3.21}
$$

3.12 Environmental Data Examples

The estimation methods discussed in this chapter are now applied to find suitable models for two data sets: the methane (CH_4) gas data shown in Table A.2 and the carbon dioxide (CO_2) data shown in Table A.3. These tables use data from the United Nations Environmental Program (UNEP 1989), recording CO_2 and CH_4 concentrations in gases trapped in the ice at the South Pole 8000 years ago.

Four nonlinear regression models were proposed by Riazoshams and Midi (2014) for these data sets. Among those considered, the exponential with intercept data set is preferred for modeling the CH_4 data because of its better convergence and the lower correlation exhibited between parameters. On the other hand, the scale exponential convex model is appropriate for the CO_2 data because, besides having smaller standard errors of parameter estimates and smaller residual standard errors, it is numerically stable. As there is a large quantity of data, going back 8000 years, there is great dispersion in the data set, so we have applied robust nonlinear regression estimation methods to create a smoother model.

Human activities are overwhelmingly the dominant contribution to the current disequilibrium of the global carbon cycle. Many researchers have attempted to explore the impact of human activity on the amount of greenhouse gases, such as CH_4 and CO_2, in the atmosphere. They have tried to find mathematical models for the changes over time and have measured the concentration of these gases trapped inside ice from thousands of years ago. UNEP (1989) reported atmospheric carbon dioxide and methane concentration data collected from cores taken from 8000-year-old ice at the South Pole.

Etheridge et al. (1998) presented the CH_4 mixing ratios from 1000 AD that were present in Antarctic ice cores, Greenland ice cores, the Antarctic firm layer, and archived air from Tasmania, Australia. Many authors have attempted to find vulnerabilities associated with CH_4 exchange. For example, Dolman et al. (2008) and also Etheridge et al. (1998) discussed modeling of CH_4 changes. Dolman et al. (2008) state that the CH_4 model is linear in the pre-industrial era and exponential in the industrial era. In recent times the rate of increase has declined. Moreover, there have also been several attempts to forecast CO_2 and CH_4 concentrations, for example Raupach et al. (2005).

Most of this research has focused on data from 1000 AD to the present day. The UNEP (1989) data set, meanwhile, goes back to 7000 BC but has high leverage values (see Figure 3.4). Riazoshams and Midi (2014) attempted to find suitable nonlinear models for CH_4 and CO_2 gas concentrations in which the high leverage values were taken into consideration. Because of the linearity of the data in the pre-industrial era, the sharp curvature in industrial era, and the high slope increase in the modern era, fitting a nonlinear model is not straightforward and some modifications of the models are required. The computational problems of these modified models are discussed in Chapter 7.

3.13   Nonlinear Models

The UNEP (1989) data for CH_4 and CO_2 concentrations collected from ice cores from the South Pole are shown in Figures 3.4 and 3.5. As can be seen from Figure 3.4, the CH_4 data contain high leverage points, so robust methods are needed to reduce their effect.

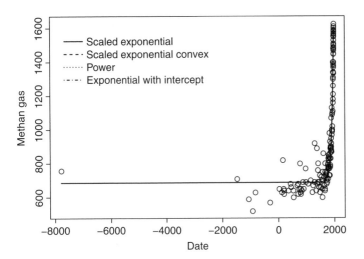

Figure 3.4 Four models fitted to methane data using the robust MM-estimator. *Source*: Riazoshams and Midi (2013). Reproduced with permission of University Putra Malaysia.

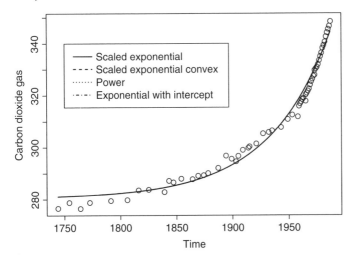

Figure 3.5 Four models fitted to carbon dioxide data using the robust MM-estimator. *Source*: Riazoshams and Midi (2013). Reproduced with permission of University Putra Malaysia.

Four nonlinear models are proposed in Equation 3.22. These models are thought to be able to describe the near-linear pre-industrial part of the data, the sharp change during the industrial era, and the linear change during the modern era with a slight decline in the rate of increase. It would be hard to fit a nonlinear model to data that behaves in this way.

To find appropriate nonlinear models, we first started to fit the data with the exponential model because of the exponent behavior of the data. Then we observed that in the exponent, location and scale parameters are needed (like p_3 and p_4 in the scaled exponential). Since the data at minus infinity are asymptotically constant, a constant parameter is added to the models. As can be seen, all four models have a constant intercept to express the horizontal asymptote at minus infinity.

Model 1: Scaled exponential $\qquad y_i = p_1 + p_2 e^{\frac{x_i - p_3}{p_4}}$ (3.22a)

Model 2: Scaled exponential convex $\qquad y_i = p_1 + e^{-(p_2 - p_3 x_i)}$ (3.22b)

Model 3: Power model $\qquad y_i = \dfrac{1}{p_1} - p_2 \cdot p_3^{x_i}$ (3.22c)

Model 4: Exponential with intercept $\qquad y_i = p_1 + e^{\frac{x_i - p_2}{p_3}}$ (3.22d)

Tables 3.1–3.4 display the parameter estimates, standard errors, correlation of parameters, and fitted correlations for the CH_4 data, with and without the high leverage points. The data are not well behaved and convergence is hard to achieve.

For the scaled exponential model 3.22a, it can be seen that the derivative with respect to p_2 is different only by a constant product of the derivative with respect to p_3. This makes the columns of the gradient matrix linearly dependent, and it is therefore singular. This fact theoretically means that the parameters are not estimable in a linear regression approximation by Taylor expansion. In this case, direct optimization using derivative-free methods, or the Levenberg–Marquardt approach in a singularity situation, can be used.

Using the robust method, the convergence is fast, with six or eight iterations required for data without and with high leverage points, respectively. The standard errors of parameters (p_2, p_2, p_3) without high leverage are $(1.116 \times 10^{-8}, 2.845 \times 10^{-4}, 4.750 \times 10^{-2}, 4.360 \times 10^{-9})$, respectively. They are larger in the presence of outliers: $(1.220 \times 10^{-7}, 2.304 \times 10^{-9}, 1.260 \times 10^{-11}, 5.539 \times 10^{-8})$. For data with high leverage points, the covariance matrix is singular; that is,

$$\hat{V}^T \hat{V} = \left\{ \begin{array}{cccc} 144 & 8.492\text{e}4 & -4.64\text{e}2 & -3.363\text{e}3 \\ & 1.365\text{e}8 & -7.458\text{e}5 & -5.570\text{e}6 \\ & & 4.078\text{e}3 & 3.046\text{e}4 \\ & & & 2.279\text{e}5 \end{array} \right\}$$

Table 3.1 The robust MM- and classical OLS estimates for the scaled exponential model, CH_4 data.

Data without high leverage points						
Parameters	OLS	Standard error	Correlation matrix			
p_1	712.04	0.14571	1	5.382×10^{-4}	5.384×10^{-4}	-0.585
p_2	0.2875	4302.42		1	0.9994	-4.260×10^{-4}
p_3	1421.5	1048576			1	-4.263×10^{-4}
p_4	70.058	0.0577				1
σ	62.656					
Correlation	0.9814					
Iterations	9					

Data with high leverage points						
Parameters	MM	Standard error*	Correlation matrix			
p_1	704.51	8.231	1	-3.7×10^{-5}	-6.65×10^{-8}	-0.58
p_2	0.44	2509.7		1	0.999	6.1×10^{-5}
p_3	1454.57	398694			1	1.21×10^{-7}
p_4	69.47	3.213				1
σ	56.63					
Correlation	0.96					
Iterations	3					

Data with high leverage points						
Parameters	OLS	Standard error	Correlation matrix			
p_1	691.3089	0.12315901	1	2.37×10^{-5}	-2.34×10^{-5}	-0.52
p_2	0.39243	3853.4942		1	1	-7.77×10^{-5}
p_3	1400.148	741455.20			1	-7.83×10^{-5}
p_4	75.50851	0.0568263				1
σ	65.66286					
Correlation	0.979283					
Iterations	13					

Data with high leverage points				
Parameters	MM	Standard error	Correlation matrix	
p_1	685.57	(*)	(*)	
p_2	0.05			
p_3	1255.45			
p_4	74.59			
σ	59.17			
Correlation	0.96			
Iterations	7			

(*) values cannot be computed.
Source: Riazoshams and Midi (2013). Reproduced with permission of University Putra Malaysia.

Table 3.2 The robust MM- and classical OLS estimates for the exponential convex model, CH_4 data.

Data without high leverage points					
Parameters	OLS	Standard error	Correlation matrix		
p_1	712.04	9.09	1	0.59	0.58
p_2	21.54	1.45		1	0.999
p_3	0.01	0.7×10^{-3}			1
σ	62.4				
Correlation	0.962				
Iterations	12				
Data with high leverage points					
Parameters	MM	Standard error	Correlation matrix		
p_1	685.55	7.267356	1	0.529	0.52
p_2	19.79	1.172492		1	0.99
p_3	0.01	0.6×10^{-3}			1
σ	59.17				
Correlation	0.96				
Iterations	50				

Source: Riazoshams and Midi (2013). Reproduced with permission of University Putra Malaysia.

with eigenvalues equal to (1.366×10^8, 4.700×10^2, 6.220×10, and -1.478×10^{-12}); the single negative eigenvalue shows that the matrix is singular. Without high leverage points, the same matrix has positive eigenvalues (1.236×10^8, 4.966×10^2, 4.479×10, and 2.615×10^{-12}). This is probably because the derivatives of the second and third parameters are linearly related and the data include high leverage points. Since the singular gradient matrix can be solved by the Levenberg–Marquardt method, we suspect that high leverage points are responsible for this problem. Furthermore, for this model the correlation between p_2 and p_3 is almost equal to 1. It is therefore appropriate to remove the parameter p_2 from this model, and this leads us to the exponential with intercept model 3.22d.

It can be seen from the results of the scaled exponential convex model in Table 3.2 that the standard errors of robust MM-estimates are lower than the OLS estimates. The correlation between p_2 and p_3 and the number of iterations is slightly higher for the MM method than for the OLS method in both situations; that is, in the presence or absence of outliers. It is important to note that the correlation between parameters is high, but not higher than in the first model.

The results for the power model 3.22c in Table 3.3 reveal that convergence is hard to achieve, as can be seen for data without high leverage points, where the

Table 3.3 The robust MM- and classical OLS estimates for the power model, CH_4 data.

Data without high leverage points					
Parameters	OLS	Standard error	Correlation matrix		
p_1	1.034×10^{-3}	5.74×10^{-5}	1	-0.720	-0.703
p_2	-1.062×10^{-3}	0.011		1	0.999
p_3	1.006153	0.005			1
σ	268.0398				
Correlation	-5.006				
Iterations	76				
Data without high leverage points					
Parameters	MM	Standard error	Correlation matrix		
p_1	1.42×10^{-3}	1.38×10^{-5}	1	-0.59	-0.58
p_2	-3.48×10^{-10}	3.92×10^{-10}		1	0.999
p_3	1.014	5.62×10^{-4}			1
σ	47.156				
Correlation	0.962				
Iterations	841				
Data with high leverage points					
Parameters	OLS	Standard error	Correlation matrix		
p_1	1.447×10^{-3}	1.69×10^{-5}	1	-0.53	-0.523
p_2	-3.47×10^{-9}	4.48×10^{-9}		1	0.999
p_3	1.013	6.61×10^{-4}			1
σ	65.430				
Correlation	0.957				
Iterations	197				
Data with high leverage points					
Parameters	MM	Standard error	Correlation matrix		
p_1	0.001458	1.54×10^{-5}	1	-0.529	-0.522
p_2	-2.53×10^{-9}	2.85×10^{-9}		1	0.999
p_3	1.013519	0.000603			1
σ	59.159891				
Correlation	0.9575862				
Iterations	195				

Source: Riazoshams and Midi (2013). Reproduced with permission of University Putra Malaysia.

Table 3.4 The robust MM- and classical OLS estimates for the exponential with intercept model, CH_4 data

Data without high leverage points					
Parameters	OLS	Standard error	Correlation matrix		
p_1	712.0400	9.09	1	0.61	−0.58
p_2	1508.850	24.20		1	−0.999
p_3	70.0600	3.60			1
σ	62.4000				
Correlation	0.961754				
Iterations	14				
Data without high leverage points					
Parameters	OLS	Standard error	Correlation matrix		
p_1	703.68	6.85	1	0.61	−0.58
p_2	1512.25	17.97		1	−0.998
p_3	69.43	2.67			1
σ	47.14				
Correlation	0.962				
Iterations	12				
Data with high leverage points					
Parameters	OLS	Standard error	Correlation matrix		
p_1	691.31	8.06	1	0.55	−0.523
p_2	1470.78	24.92		1	−0.998
p_3	75.51	3.72			1
σ	65.43				
Correlation	0.957242				
Iterations	15				
Data with high leverage points					
Parameters	OLS	Standard error	Correlation matrix		
p_1	684.77	6.06	1	0.548	−0.522
p_2	1476.49	18.45		1	−0.998
p_3	74.55	2.75			1
σ	49.37				
Correlation	0.958				
Iterations	8				

Source: Riazoshams and Midi (2013). Reproduced with permission of University Putra Malaysia.

robust method needs 841 iterations. The OLS gives a poor fit, with the wrong value of fitted model correlation at −5.006, which is due to the wrong parameter estimates and a large value for the residual standard error for the model 3.22c, which is possibly due to rounding errors (see Table 3.3). Figure 3.4 clearly reveals the poor fit of the model. The robust method works better for both cases and is more reliable, although some far-away points still can be seen after the high leverage points have been removed from the data. It is important to note that the correlation between parameters p_2 and p_3 is still high.

Table 3.4 shows the estimates the exponential with intercept model 3.22d (p_1 is called an intercept). As explained previously for the scaled exponential model, parameters p_2 and p_3 have an almost linear relation, which encourages us to remove p_2, leading to model 3.22d. The intercept, p_1, is the limit value of the data at ($-\infty$); that is, the amount of methane gas in preindustrial times. The correlation between parameters in the worst case is better than in other models.

In the presence of high leverage points, the robust estimates of the residual standard errors of the first three models are very close:

- 59.17 for model 3.22a (Table 3.1)
- 59.17 for model 3.22b (Table 3.2)
- 59.16 for model 3.22c (Table 3.3).

However, the values are higher than for model 3.22d. It can be seen that the power and scaled exponential convex models have poor convergence and higher correlations between parameters. Figure 3.4 shows the robust MM fits of the four models in the presence of high leverage points, which suggests a closer fit to the data. However, the exponential with intercept model is preferred because it has the lowest value of the residual standard error, needs fewer iterations, and is a better fit than other models.

The plots in Figure 3.4 suggest the possibility of heteroscedastic errors: the variance of the errors is decreasing in a systematic manner with the increase in x values. This problem is not considered in the analysis here since the degree of heteroscedastic error seems to be small.

3.14 Carbon Dioxide Data

UNEP (1989) presented CO_2 data collected from the same source as the CH_4 gas data. The data do not have outliers and so it is easier to fit the models. Tables 3.5–3.8 show the parameter estimates, standard errors, correlation of parameters, and fitted correlations for the CO_2 data.

Let us first focus on the scaled exponential model of Table 3.5. Similar to the CH_4 data, the OLS estimates cannot be computed by the modified Newton method and so the Levenberg–Marquardt method is used instead.

Table 3.5 The robust MM- and classical OLS estimates for the scaled exponential model, CO_2 data.

Data without high leverage points						
Parameters	OLS	Standard error	Correlation matrix			
p_1	280.3578	(*)	(*)			
p_2	0.2329					
p_3	1681.586					
p_4	54.3729					
σ	2.6474361					
Correlation	0.9926157					
Iterations	6					
Data with high leverage points						
Parameters	OLS	Standard error	Correlation matrix			
p_1	280.3468	5.449×10^{-4}	1	-1.72×10^{-4}	-4.23×10^{-7}	-8.1678×10^{-1}
p_2	0.1382	0.115375		1	1	2.0462×10^{-4}
p_3	1652.462	45.494549			1	5.1831×10^{-7}
p_4	54.4838	0.0014943				1
σ	3.054331					
Correlation	0.9850827					
Iterations	6					

Source: Riazoshams and Midi (2013). Reproduced with permission of University Putra Malaysia.

Table 3.6 The robust MM- and classical OLS estimates for the scaled exponential convex model, CO_2 data

Data without high leverage points					
Parameters	OLS	Standard error	Correlation matrix		
p_1	280.358	1.022	1	0.820	0.8162
p_2	32.38431	1.88758		1	0.999
p_3	0.01839	9.480×10^{-4}			1
σ	2.624904				
Correlation	0.985				
Iterations	10				
Data with high leverage points					
Parameters	OLS	Standard error	Correlation matrix		
p_1	280.346	2.79×10^{-4}	1	0.821	0.8168
p_2	32.30756	5.1315×10^{-4}		1	0.999
p_3	0.01835	2.5772×10^{-7}			1
σ	3.054396				
Correlation	0.985				
Iterations	9				

Source: Riazoshams and Midi (2013). Reproduced with permission of University Putra Malaysia.

Table 3.7 The robust MM- and classical OLS estimates for the power model, CO_2 data

Data without high leverage points					
Parameters	OLS	Standard error	Correlation matrix		
p_1	3.632×10^{-3}	6.99×10^{-61}	1	−0.885	−0.881
p_2	-5.5×10^{-11}	3.36×10^{-11}		1	0.999
p_3	1.0140958	3.079×10^{-4}			1
σ	3.0325				
Correlation	0.990				
Iterations	100				
Data with high leverage points					
Parameters	OLS	Standard error	Correlation matrix		
p_1	0.0036	5.8018×10^{-5}	1	−0.886	−0.882
p_2	-6.31×10^{-11}	0.000		1	0.999
p_3	1.014	0.003			1
σ	3.0782				
Correlation	0.9732				
Iterations	101				

Source: Riazoshams and Midi (2013). Reproduced with permission of University Putra Malaysia.

Table 3.8 The robust MM- and classical OLS estimates for the exponential with intercept model, CO_2 data

Data without high leverage points					
Parameters	OLS	Standard error	Correlation matrix		
p_1	280.3581	1.022	1	0.849	−0.816
p_2	1760.812	1.1896×10^1		1	−0.997
p_3	54.37217	2.803			1
σ	2.625				
Correlation	0.985				
Iterations	18				
Data with high leverage points					
Parameters	OLS	Standard error	Correlation matrix		
p_1	280.52	1.9584×10^{-6}	1	0.848	−0.815
p_2	1761.76	2.278×10^{-5}		1	−0.997
p_3	54.17	5.370×10^{-6}			1
σ	2.22				
Correlation	0.985				
Iterations	8				

Source: Riazoshams and Midi (2013). Reproduced with permission of University Putra Malaysia.

The correlation of the fit is high, with convergence achieved after six iterations, but the correlation between p_2 and p_3 is one. Similar to the results for the CH_4 data, this leads us to the exponential with intercept model.

Table 3.6 reveals that the MM-estimates are fairly close to the OLS estimates for the scaled exponential convex model 3.22b, since the data do not have outliers. It is interesting to note that the standard errors of the MM-estimates are slightly smaller than the OLS estimates but have higher residual standard errors. In this situation, the OLS method is preferred.

Table 3.7 shows the results for the power model. Similar to the CH_4 data, convergence is difficult to achieve for this model. It can be seen that the OLS and MM methods require 100 and 101 iterations, respectively. The parameter estimates seem to be very small and the model needs to be rescaled.

It can be seen from Table 3.8 that the OLS method needs more iterations than the MM method. We encountered computational problems in both the OLS and MM methods. The models were highly sensitive to the initial values. The results for model 3.22d are similar to the results for model 3.22c, but model 3.22b is preferred as it has fewer computational problems. However, judging from the residual standard errors and the standard errors of the parameter estimates, both can be recommended for this data.

Figure 3.5 displays the fits of the four models using the robust MM method. The plot suggests there is not a big difference between the fitted lines. After examining the residual plots, we can see that the errors are autocorrelated, but no further analysis is considered to remedy this problem.

3.15 Conclusion

Four models are proposed and fitted to the CH_4 data presented by UNEP (1989). Three out of the four models are feasible for use, but the exponential with intercept model is preferred due to its better convergence and the lower correlation between parameters. The intercept is the preindustrial limit value from the robust fit for the exponential with intercept model. When time tends to $-\infty$ the model tends to parameter $p_1 = 648.77$, which is the value of the methane concentration 7000 years ago.

The robust MM-estimator can balance the curve between preindustrial and more recent times better than the OLS estimator. This helps to reduce the effect of measurement errors (the data are collected by measuring the concentration of gases trapped in ice at the poles 8000 years ago, and so it is not guaranteed that they are free of errors). Based on the UNEP report (1989), only good values are presented in the graphs and no model has been proposed for the data. Riazoshams and Midi (2014) is the first attempt to model it. Environmental scientists might be able to suggest further parameters that could be included.

For the CO_2 data, models 3.22b and 3.22d fit reasonably well. The standard error of the parameter estimates and the residual standard errors of model 3.22d are fairly closed to those for model 3.22b. Nonetheless, fitting for model 3.22d posed certain computational problems. On the other hand, the second model is numerically stable. In this respect, model 3.22b is preferred for the CO_2 data.

4

Heteroscedastic Variance

Nonlinear regression as defined in previous chapters assumed that errors are identically independent distributed (i.i.d.) with constant (homogeneous) variance. This chapter discuss the cases when the variance of the error is not homogeneous. The error variance is called heteroscedastic when it follows a functional relation with the predictor. In general it is a nonlinear function with some unknown parameters. Estimating heteroscedastic variance function parameters and fitting a nonlinear regression model with heteroscedastic variance is the subject of this chapter.

Heteroscedasticity of error variance is seen in real-life examples for several reasons. A classic example (without outliers) is the chicken weight data presented by Riazoshams and Miri (2005) (Table A.1). The authors showed that the variance structure of the data is a power function of the nonlinear regression model $f(x_i; \theta)$. Although the variance can be a general function form of the predictor variable x_i, in many applications it is observed the variance can be written as a function of the nonlinear regression model $f(x_i; \theta)$ (an example will appear later on, in Figure 4.1).

As can be seen from the chicken weight data, independence or homogeneity assumptions about the regression residuals can be broken due to the natural behavior of data. In fact, in real-life examples these null conditions are natural, and analysts must take them into consideration. If they fail to do so, further inferences, especially outlier detection, might be inappropriate.

In the nonlinear regression

$$y_i = f(x_i; \theta) + \varepsilon_i, \ i = 1, \ldots, n$$

the error variance is generally assumed to be constant; let us say $(var(\varepsilon_i) = \sigma^2)$. The heterogeneity or heteroscedasticity of variance refers to the case when the variance is not constant. In such a case, the variance of errors can be written as:

$$Var(\varepsilon_i) = \sigma_i^2 \tag{4.1}$$

In real-life examples several cases of heterogeneity can occur.

Robust Nonlinear Regression: with Applications using R, First Edition.
Hossein Riazoshams, Habshah Midi, and Gebrenegus Ghilagaber.
© 2019 John Wiley & Sons Ltd. Published 2019 by John Wiley & Sons Ltd.
Companion website: www.wiley.com/go/riazoshams/robustnonlinearregression

Case 1: The variance can be a known quantity for each single data point but with different values:

$$\sigma_i^2 = w_i^2 \tag{4.2}$$

where w_i are known values called weights. Such weights can obey a certain pattern that can be represented as a function of the predictors.

Case 2: The error variance or weight can have a certain pattern that can be represented as a function (denoted by G) of the predictors or response:

$$\sigma_i^2 = G(x_i, y_i) \tag{4.3}$$

which in inference and computation is similar to the first case, and does not require any mathematical extension.

Case 3: The error variance can be a general known parametric function of predictors

$$\sigma_i^2(\tau) = G(x_i; \tau) \tag{4.4}$$

with unknown variance model parameters $\tau \in \mathfrak{R}^q$.

Case 4: In some applications the error variance can be written as a function of the nonlinear regression model as

$$\sigma_i^2(\tau) = G(f(x_i; \theta); \tau) \tag{4.5}$$

Theoretically this case is similar to the third case but in computation can be different. Therefore, in the computing part of this book we will define the objects in a different way.

It is more practical to assume that the model includes a fixed scale parameter representing the total dispersion. That means that equations (4.2)–(4.5) might include a constant variance value σ^2. For example, the general variance model G can be expressed as $G(x_i; \tau) = \sigma^2 g(x_i; \lambda)$, for real valued functions g and $\tau = (\sigma, \lambda) \in \mathfrak{R}^q$. The four cases can be rewritten as:

Case 1:	$\sigma_i^2 = \sigma^2 w_i^2$	(4.6)
Case 2:	$\sigma_i^2 = \sigma^2 g(x_i, y_i)$	(4.7)
Case 3:	$\sigma_i^2(\tau) = \sigma^2 g(x_i; \lambda)$	(4.8)
Case 4:	$\sigma_i^2(\tau) = \sigma^2 g(f(x_i; \theta); \lambda)$	(4.9)

4.1 Definitions and Notations

For simplicity, we are using the following notation. Index i is used to show the ith data point value or a function value at point x_i. Bold symbols are used for vectors of functions. A dot indicates a gradient or gradient matrix, a double-dot indicates a Hessian matrix, which might be a three-dimensional array in which the first dimension includes the data points of the Hessian matrix.

The definitions used are as follows:

- Nonlinear regression function model at point i: $\eta_i(\theta) = f(x_i; \theta)$
- Vector of nonlinear regression function model: $\boldsymbol{\eta}(\theta) = [\eta_1, \dots, \eta_n]^T$
- Variance function model at point i: $G_i(\tau) = G(x_i; \tau)$
- Vector of variance function model at point i: $\mathbf{G}(\tau) = [G_1, \dots, G_n]$
- g function at point i: $g_i(\lambda) = g(x_i; \lambda)$
- Vector of g functions: $\mathbf{g}(\lambda) = [g_1, \dots, g_n]^T$
- Gradient of nonlinear regression function model: $\dot{\eta}_i(\theta) = \left[\frac{\partial f(x_i; \theta)}{\partial \theta}\right]_{p \times 1}$
- Gradient matrix of nonlinear regression function model: $\dot{\boldsymbol{\eta}}(\theta) = [\dot{\eta}_i^T(\theta)]_{n \times p}$
- Hessian matrix of nonlinear regression function model: $\ddot{\eta}_i(\theta) = \left[\frac{\partial^2 f(x_i; \theta)}{\partial \theta^T \partial \theta}\right]_{p \times p}$
- Array of Hessian matrix of nonlinear model function: $\ddot{\boldsymbol{\eta}}(\theta) = [\ddot{\eta}_i(\theta)]_{n \times p \times p}$
- Gradient of variance function model: $\dot{G}_i(\theta) = \left[\frac{\partial G(x_i; \tau)}{\partial \tau}\right]_{p \times 1}$
- Gradient matrix of variance function model: $\dot{\mathbf{G}}(\tau) = [\dot{G}_i^T(\theta)]_{n \times p}$
- Hessian matrix of variance function model: $\ddot{G}_i(\tau) = \left[\frac{\partial^2(x_i; \tau)}{\partial \tau^T \partial \tau}\right]_{q \times q}$
- Array of Hessian matrix of nonlinear model function: $\ddot{\mathbf{G}}(\tau) = [\ddot{G}(\tau)]_{n \times q \times q}$
- Analogously we can define $\dot{g}_i$, $\dot{\mathbf{g}}$, $\ddot{g}_i$, $\ddot{\mathbf{g}}$
- Define errors as $r_i(\theta) = y_i - f(x_i; \theta)$; analogously we can define $\mathbf{r}_n(\theta)$, $\dot{r}_i(\theta)$, $\dot{\mathbf{r}}_{n \times p}(\theta)$, $\ddot{r}_i(\theta)$, $\ddot{\mathbf{r}}_{n \times p \times p}(\theta)$

4.2 Weighted Regression for the Nonparametric Variance Model

The weighted cases 1 and 2 for the variance model ((4.2) and (4.3)) can be solved by weighted regression. In this case the variance values are known, which means there is no parameter in the variance form. In contrast, there are unknown parameters in the error variance model. In such a case, we start with estimating the model function parameter θ, then estimate the variance model parameter τ, then compute the final values of θ using weighted regression.

The weighted regression is in fact a special case of the generalized least squares method covered in Section 2.3.3. Assuming the variance structure is known, as in cases 1 or 2 above, then the nonlinear regression model can be written as:

$$y_i = f(x_i; \theta) + w_i \xi_i, \quad i = 1, \dots, n$$

in which the errors ξ_i are i.i.d. with constant variance $Var(\xi_i) = \sigma^2$. The least squares estimate is the answer to the minimization of the sum of square errors equation given by:

$$S(\theta) = \sum_{i=1}^{n} w_i^2 (y_i - f(x_i; \theta))^2.$$

This is equivalent to the generalized sum of squares estimate (2.17):

$$\hat{\theta} = \arg \min_{\theta} \ [\mathbf{y} - \boldsymbol{\eta}]^T V [\mathbf{y} - \boldsymbol{\eta}] \tag{4.10}$$

where

$$Var(\varepsilon) = \sigma^2 \text{diag}[w_i^2] \tag{4.11}$$

and

$$V = \text{diag}[w_i^2] \tag{4.12}$$

The more general case of the heterogeneous variance form (4.8), when the variance model parameter is known, can be represented similarly as:

$$Var(\varepsilon) = diag[G(x_i; \sigma^2, \lambda)] = \sigma^2 \begin{bmatrix} g(x_1; \lambda) & 0 & \cdots & 0 \\ 0 & g(x_2; \lambda) & \cdots & 0 \\ \vdots & \vdots & \vdots & \vdots \\ 0 & 0 & \cdots & g(x_n; \lambda) \end{bmatrix}$$

$$\tag{4.13}$$

with $V = \text{diag}[g(x_i; \lambda)]$. Let the Cholesky decomposition of V be $V = U^T U$, $U = \text{diag}[w_i]$. Using inverse transpose $R = (U^T)^{-1} = \text{diag}[\frac{1}{w_i}]$, the model can be transformed into a homogeneous variance model with constant variance (2.18), and all the computations can be followed from Section 2.3.3.

The weighted regression and consequently the generalized least squares method can be used for the other cases where the variance model is a function of an unknown parameter. If the variance model parameter is known, then the variance values can be computed and the nonlinear function parameter can be estimated by generalized least squares. In this regard, there are a few methods to estimate the parameter τ, and these are discussed in the rest of this chapter.

4.3   Maximum Likelihood Estimates

In the general parametric variance form (case 3), both parameters of the nonlinear model θ and variance model parameter τ must be estimated. If the errors follow a normal distribution, the maximum likelihood estimates of parameters θ and τ can be computed simultaneously by maximizing (minimizing) the (minus) generalized form of the log-likelihood function in Equation (2.7).

Assume that the variance model function defined in parameter space $\tau \in \mathfrak{R}^q$ follows the general parametric form given by (4.8):

$$G(\tau) = \sigma_i^2(\tau) = \sigma^2 g(x_i; \lambda) \tag{4.14}$$

which transfers the space of predictors $x_i \in \mathfrak{R}^k$ into the parameter space of $\tau \in \mathfrak{R}^q$. The log-likelihood function is:

$$\ell(\theta, \tau) = -\sum_{i=1}^{n} \left\{ \log[\sigma^2 g(x_i; \lambda)] + \frac{[y_i - f(x_i; \theta)]^2}{\sigma^2 g(x_i; \lambda)} \right\} \tag{4.15}$$

Then the MLE of parameters $(\theta, \lambda, \sigma)$ is calculated from the simultaneous matrix form equations:

$$\begin{cases} \dfrac{\partial \ell}{\partial \theta} = \dfrac{1}{\sigma^2} \dot{\eta}(\theta) \left[\dfrac{\mathbf{r}(\theta)}{\mathbf{g}(\lambda)} \right] = 0 & (4.16a) \\[4mm] \dfrac{\partial \ell}{\partial \lambda} = -\dot{\mathbf{g}}^T(\lambda) \left[\dfrac{1}{\mathbf{g}(\lambda)} - \dfrac{\mathbf{r}^2(\theta)}{\sigma^2 \mathbf{g}^2(\lambda)} \right] = 0 & (4.16b) \\[4mm] \dfrac{\partial \ell}{\partial \sigma^2} = -\left[\dfrac{1}{\sigma^2} - \dfrac{\mathbf{r}^2(\theta)}{\sigma^4 \mathbf{g}(\lambda)} \right] \mathbf{1}_n^T = 0 \Rightarrow \hat{\sigma}^2 = \dfrac{1}{n} \sum_{i=1}^{n} \dfrac{r_i^2}{g_i} & (4.16c) \end{cases}$$

The second derivative is given by:

$$\frac{\partial^2 \ell}{\partial(\theta, \lambda, \sigma^2)^T \partial(\theta, \lambda, \sigma^2)} = \begin{bmatrix} \ell_{11} & \ell_{12} & \ell_{13} \\ " & \ell_{22} & \ell_{23} \\ " & " & \ell_{33} \end{bmatrix}$$

in which

$$\ell_{11} = \frac{\partial^2 \ell}{\partial \theta^T \partial \theta} = \ddot{\eta}(\theta) \otimes \left[\frac{\mathbf{r}}{\sigma^2 \mathbf{g}(\lambda)} \right] - \left[\left[\frac{1}{\sigma^2 \mathbf{g}(\lambda)} \right] \circ \dot{\eta}_{\bullet j}(\theta) \right]^T \dot{\eta}(\theta)$$

$$\ell_{12} = \frac{\partial^2 \ell}{\partial \theta^T \partial \lambda} = -\left[\frac{\mathbf{r}(\theta)}{\sigma^2 \mathbf{g}^2(\lambda)} \circ \dot{\eta}_{\bullet j}(\theta) \right]^T \dot{\mathbf{g}}(\lambda)$$

$$\ell_{13} = \frac{\partial^2 \ell}{\partial \theta^T \partial \sigma^2} = -\frac{1}{\sigma^4} \dot{\eta}(\theta) \left[\frac{\mathbf{r}(\theta)}{\mathbf{g}(\lambda)} \right]$$

$$\ell_{22} = \frac{\partial^2 \ell}{\partial \lambda^T \lambda} = -\ddot{g}(\lambda) \otimes \left[\frac{1}{g(\lambda)}\right] - \left[\left[\frac{1}{g(\lambda)}\right] \circ \dot{g}_{\bullet j}(\lambda)\right]^T \dot{g}(\lambda)$$

$$\ell_{23} = \frac{\partial^2 \ell}{\partial \lambda \partial \sigma^2} = -\frac{1}{\sigma^4}\left[\frac{r^2(\theta)}{g^2(\lambda)}\right]\dot{g}(\lambda)$$

$$\ell_{33} = \frac{\partial^2 \ell}{\partial^2 \sigma^2} = \frac{n}{\sigma^4} + \frac{2}{\sigma^2}\left[\frac{1}{g(\lambda)}\right]^T r^2(\theta)$$

Therefore, since the expectation of errors is zero $(E(r(\theta)) = 0)$, we can conclude that the Fisher information is equal to:

$$I_{(p+q+1)\times(p+q+1)} = \begin{bmatrix} -\left[\left[\frac{1}{\sigma^2 g(\lambda)}\right]\circ\dot{\eta}_{\bullet j}(\theta)\right]^T \dot{\eta}(\theta) & 0 & 0 \\ 0 & -\ddot{g}(\lambda)\otimes\left[\frac{1}{g(\lambda)}\right] & 0 \\ 0 & 0 & \frac{n}{\sigma^4} \end{bmatrix}$$

Given the estimates of the nonlinear model function parameter $\hat{\theta}$ and the variance model parameters $(\hat{\sigma}, \hat{\lambda})$, the parameter confidence intervals and confidence interval for response at point X_0 can be extended intuitively from linear regression techniques (Table 1.1) by linear approximation of the nonlinear model using the gradient matrix. Therefore, the design matrix is replaced by the gradient matrix:

$$\hat{Y}_0 \in [\eta_0 \pm \hat{\sigma}[g(X_0, \hat{\lambda}) + \dot{\eta}_0(\dot{\eta}^T \dot{\eta})\dot{\eta}_0]^{(1/2)}]$$

in which all function models $(\eta, g, \dot{\eta})$ are computed at the estimated parameter values $(\hat{\theta}, \hat{\tau})$, $\eta_0 = f(X_0, \hat{\theta})$.

4.4 Variance Modeling and Estimation

Identifying the nonlinear regression model and the structure of the error variance is not straightforward. In general, all models are wrong and we want to find the best from several possible models. This is the problem of interest for both the nonlinear regression model and the variance model. The procedure is called model selection. There are several criteria for selecting the best model, such as the Akaike information criterion (AIC), the Bayes information criterion (BIC), and the cross-validation criterion (CV) (see Bunke et al. (1999)).

In order to estimate the parameter vector τ of the variance model in (4.4), Bunke et al. (1998) first calculated the sample variance and then used the chi-square distribution of the observed variances. If the ith observation x_i is repeated n_i times with response values $y_{ij}, j = 1, \dots, n_i$, then the observation variances σ_i^2 can be estimated by intra-sample variance estimates:

$$s_i^2 = \frac{1}{n_i - 1}\sum_{i=1}^{n_i}(y_{ij} - \bar{y}_i)^2, \quad \bar{y}_i = \frac{1}{n_i}\sum_{i=1}^{n_i}y_{ij} \tag{4.17}$$

If there are enough replications (n_i is relatively large), an improved variance estimate may be possible. Bunke et al. (1998, 1999) defined observed sample variances $z_1, \ldots, z_n$ as in (4.18), such that the expectation of z_i is approximately equal to σ_i^2:

$$z_i = \begin{cases} s_i^2 & n_i \geq 2 \\ \frac{1}{2}|e_{i+1} - e_i| & n_i = 1 \text{(no replicate)} \end{cases} \tag{4.18}$$

where the residuals e_i are defined by

$$e_i = y_{i1} - f(x_i, \hat{\theta}) \tag{4.19}$$

For $n_i = 2$, Bunke et al. (1999) proposed repeating the data points appropriately, for reasons of efficiency. The sample variances τ_i are not robust. Simulations show that outliers that are influential in nonlinear regression parameter estimates are not necessarily influential in calculating the sample variance. This observation persuades us to not only make the estimate of the parameters of the models robust for both the regression function and the variance, but also to replace the sample variance of the data by a robust scale estimate. This inspires us to replace the classical variance estimator with a more robust measure.

Under the assumption of normality for errors, the statistics $\max(1, n_i - 1)z_i/\sigma_i^2$ are approximately chi-square distributed, $\chi^2_{\max(1,n_i-1)}$. Bunke et al. (1998, 1999) proposed fitting the alternative variance model $\tilde{H}(x; \sigma^2, \lambda) = \max(H(x; \sigma^2, \lambda), \hat{\sigma}_0^2)$ to the observations z_i by maximizing the logarithmic χ^2 pseudo-likelihood:

$$\begin{aligned} \ell(\sigma^2, \lambda) = &- \sum_{i;n_i \geq 2} (n_i - 1)\{\log(\tilde{H}(x_i; \sigma^2, \lambda)) + s_i^2/\tilde{H}(x_i; \sigma^2, \lambda)\} \\ &- \sum_{i;n_i=1} \{\log(\tilde{H}(x_i; \sigma^2, \lambda)) + z_i/\tilde{H}(x_i; \sigma^2, \lambda)\} \end{aligned} \tag{4.20}$$

Unfortunately, some variance models have the disadvantage of possibly leading to negative estimates $\hat{\sigma}_i^2 = G(x_i; \hat{\tau})$ for some design points of x_i. In such cases, the negative (and also very small) values of $G(x_i; \hat{\tau})$ will be replaced by some fixed small positive value, for example $\hat{\sigma}_0^2 = 0.1\hat{\sigma}^2$, where $\hat{\sigma}^2$ is the homogeneous variance estimate given by

$$\hat{\sigma}^2 = \frac{1}{n-q}\left\{ \sum_{i:n_i \geq 2}\sum_{j=1}^{n_i} (y_{ij} - \bar{y}_i)^2 + \sum_{i:n_i=1} z_i \right\}$$

For estimating both the model parameters θ and variance model parameter t, we use the classical multistage estimate (CME). First, the classical parameter estimates of θ and τ are combined; then a generalized least squares technique is applied. The CME procedure is defined as follows:

Algorithm 4.1 Classical multistage estimate for heteroscedastic nonlinear regression.

Stage 1: Compute the least squares estimate of θ using Equation 2.12.

Stage 2: Use the parameter estimate of θ in stage 1 to compute the sample variances z_i using Equation (4.18).

Stage 3: Use the parameter estimate of θ in stage 1 and the sample variances computed in stage 2 to estimate the parameter τ by maximizing the chi-square pseudo likelihood function in Equation 4.20.

Stage 4.1: Predict the variance function $G\left(\hat{\sigma}_i^2 = G(x_i; \hat{\tau})\right)$ using the estimate of $\hat{\tau}$ obtained in stage 3.

Stage 4.2: Calculate the estimate of the matrix V of ε in Equation 4.11 (denoted by $\text{diag}[H(x_i; \hat{\tau})] = \hat{\sigma}^2 \hat{V}$).

Stage 4.3: Estimate the parameter θ ($\hat{\theta}_{CME}$) by the feasible generalized least squares methods in Equation 2.19, where V is replaced by the estimate $\hat{V}$ from stage 4.2.

4.5 Robust Multistage Estimate

Outliers not only have masking and swamping effects on parameter estimates but also on the heteroscedasticity of variance. In other words, outliers can mask the heteroscedasticity of the data (see Midi and Jafaar (2004)), or swamp variance homogeneity, making it appear heteroscedastic.

It is important to point out that since each step of the CME technique is based on classical estimators, it is not robust. We therefore need a more robust method to replace it. The simplest solution is to identify and replace the nonrobust stages with an equivalent robust form. To this end, the parameter estimate q, in stages 1 and 4.3, is replaced by the MM-estimates and generalized MM-estimates, respectively. The parameter estimate of τ is obtained by maximizing (or minimizing) a robustified form of the (minus) chi-squared likelihood function (4.20). Since the sample variance z_i in stage 2 is sensitive to outliers, we define its robust form as follows:

$$
z_i = \begin{cases} \text{MAD}^2(y_{ij}, j = 1, \ldots, n_i) & n_i \geq 2 \\ \frac{1}{2}|e_{i+1} - e_i| & n_i = 1 \text{ (no replicate)} \end{cases} \tag{4.21}
$$

where the median absolute deviance (MAD) is a robust scale estimate. The MAD is calculated over the response values y_{ij} that correspond to point x_i, which is repeated n_i times. This book assumes the data-replicated case $n_i \geq 2$; the nonreplicated case is still under study. The normalized MAD is preferable; that is, $MAD(x) = Median\{|x - Median(x)|\}/0.6745$ (see Maronna et al. (2006)).

The terms

$$\sum_{i;n_i \geq 2} (n_i - 1)s_i^2 / \tilde{H}(x_i; \sigma^2, \lambda)$$

and

$$\sum_{i;n_i = 1} z_i / \tilde{H}(x_i; \sigma^2, \lambda)$$

inside the chi-square pseudo-likelihood function formula (2.19) are sensitive to outliers. On the other hand,

$$\sum_{i;n_i \geq 2} (n_i - 1) \log(\tilde{H}(x_i; \sigma^2, \lambda))$$

and

$$\sum_{i;n_i = 1} \log(\tilde{H}(x_i; \sigma^2, \lambda))$$

are not sensitive to outliers. For this reason, we replace the second term in the summation notation by a ρ function, just like in the M-estimator. In other words, the robust M-estimate of parameter $\tau = (\sigma^2, \lambda) \in \mathfrak{R}^q$ is obtained by maximizing the robustified chi-square log-likelihood function (or minimizing the negative of the log likelihood).

$$\ell(\sigma^2, \lambda) = - \sum_{i;n_i \geq 2} (n_i - 1)\{\log(\tilde{H}(x_i; \sigma^2, \lambda)) + \rho[s_i / \sqrt{\tilde{H}(x_i; \sigma^2, \lambda)}]\} \quad (4.22)$$

The following algorithm summarizes the proposed robust multistage algorithm for estimating the parameter function model and the variance model parameter. The proposed method (Riazoshams and Midi 2016) is called the robust multistage estimator (RME).

4.6   Least Squares Estimate of Variance Parameters

Numeric computation of the optimization problem in Equation 4.22 is difficult, especially for big data sets. In order to find a numerically easier method, the most straightforward approach is to use the least squares estimators and split the parameter estimates of θ and τ into separate consecutive stages. Riazoshams and Midi (2016) introduced a robust form of the least squares estimates for regression between sample variance and residuals, studied the properties of the estimate, and performed a comparison of the RMEs.

The sample variance z_i in 4.21 is an idea concerning the response variable of the variance model; that is, the variance model function $G(x_i; \hat{\tau})$ can be considered as a nonlinear pattern of the sample variance z_i. That results in a

Algorithm 4.2 Robust multistage estimator for heteroscedastic nonlinear regression.

Stage 1: Compute the MM-estimate of θ (Algorithm 3.1).

Stage 2: Use the stage 1 estimate to compute the robust sample variances z_i using Equation 4.21.

Stage 3: Use the parameter estimate of θ in stage 1 and sample variances computed in stage 2 for estimating the parameter τ by maximizing the robust chi-squared likelihood function (4.22).

Stage 4.1: Compute the variance function G ($\hat{\sigma}_i^2 = G(x_i; \hat{\tau})$) using the estimate of $\hat{\tau}$ obtained in stage 3.

Stage 4.2: Calculate the estimate of the covariance matrix of ε in Equation 4.13 (denoted by $diag[G(x_i; \hat{\tau})] = \hat{\sigma}^2 \hat{V}$).

Stage 4.3: Estimate the parameter θ ($\hat{\theta}_{RME}$) by the feasible generalized MM-estimator method discussed in Chapter 3 (see (3.21)) using the estimate of matrix V ($\hat{V}$) from stage 4.2.

nonlinear regression model to be fitted:

$$z = G(x; \tau) + \xi \tag{4.23}$$

where ξ is the vector of errors and $\tau = (\sigma^2, \lambda) \in \mathfrak{R}^q$ is an unknown parameter of the variance model function $G(x; \tau)$, which is written $\sigma^2 g(x_i, \lambda)$ (Equation 4.8). We are trying to avoid a nested nonlinear model by reweighting the least squares for fitting the model in Equation 4.23. For this purpose, the variance of errors ξ can be computed from the chi-square approximation distribution of z_i. As mentioned, $\max(1, n_i - 1)z_i/\sigma_i^2 \sim \chi^2_{\max(1, n_i = 1)}$. Let $w_i = \max(1, n_i - 1)$. Then

$$w_i z_i / \sigma_i^2 \sim \chi^2_{(w_i)}$$

with variance given by

$$V(z_i) = 2\sigma_i^4 / w_i \tag{4.24}$$

so the error of the nonlinear regression (4.23) has a chi-square distribution with nonconstant variance. Nevertheless, there are two useful properties that allow us to use the iteratively reweighted least squares approach. First, the expectation of this statistic is equal to the variance model function because

$$E(w_i z_i / \sigma_i^2) = w_i$$
$$w_i E(z_i) / \sigma_i^2 = w_i$$
$$E(z_i) = \sigma_i^2$$

Second, the variance of the errors ξ, for the fixed value of τ, is a known value and not related to other parameters. Therefore, the iteratively reweighted least squares approach can be applied. In fact, the chi-square pseudo likelihood (4.22) is the likelihood of the nonlinear model 4.23. To derive the weighted least squares form, if we write the nonlinear regression (4.23) in a non-matrix form and the variance function model in the form $\sigma^2 g(x_i, \lambda)$, we have

$$z_i = G(x_i; \tau) + \sqrt{2}G(x_i; \tau)/\sqrt{w_i}\zeta_i \tag{4.25}$$

where ζ_i is a random error with $E(\zeta_i) = 0$; $Var(\zeta_i) = 1$. In this case, if we consider the weighted least squares method, the weight is not constant and depends on an unknown parameter. Carroll and Ruppert (1988) discussed iteratively reweighted least squares in the case of a known variance function, and generalized the least squares algorithm to the case of an *unknown* variance function. By a change of variables we have

$$\frac{\sqrt{w_i}}{\sqrt{2}g(x_i; \lambda)}z_i = \frac{\sigma^2}{\sqrt{2/w_i}} + \sigma^2\zeta_i$$

The weight is determined so that the resulting model has a constant variance; that is,

$$V\left(\frac{\sqrt{w_i}}{\sqrt{2}g(x_i; \lambda)}z_i\right) = \frac{w_i}{2g^2(x_i; \lambda)}V(z_i)$$

$$= \frac{w_i}{2g^2(x_i; \lambda)} \times \frac{2\sigma^4 g^2(x_i; \lambda)}{w_i}$$

and the ordinary least squares (OLS) can be written as:

$$\hat{\tau} = \arg\min_\tau \sum_i \left(\frac{\sqrt{w_i}}{\sqrt{2}g(x_i; \tau)}z_i - \frac{\sigma^2}{\sqrt{2/w_i}}\right)^2$$

To simplify the computation we can use the reweighted least squares approach. Using the following form

$$\frac{z_i}{\sqrt{2/w_i}g(x_i; \tau)} = \frac{G(x_i; \tau)}{\sqrt{2/w_i}g(x_i; \tau)} + \sigma^2\zeta_i \tag{4.26}$$

the repeating weight can be considered to be $\sqrt{2/w_i}g(x_i; \tau)$. Since $E(z_i) = \sigma_i^2$ we use the direct least squares form. As Carroll and Ruppert (1988) suggest, we use two iterations in the reweighting procedure. For the first iteration, the convergence precision can be assigned with the usual numerical methods, but in the second iteration the convergence rate is reduced five times. Because, in the first iteration, the calculated value of τ is close to the optimum point, the amount of change in the τ values within the nearby iteration of the second stage is very small. Thus, a lower value of convergence rate is sufficient. This avoids a rounding error due to the closeness of the iterated sequences. Another reason for this is outlined in the second iteration, where, if the convergence precision is assigned in the first iteration, the proximity of the computed sequence requires more iterations to achieve the required convergence. We observed in some cases that the number of iterations will be more than the default iteration limit and wrongly produce nonconvergence iteration errors. The generalized least squares estimate can be written in the form:

$$\tau_{OLS} = \arg \min_{\tau}(z - G(x; \tau))^T \times diag[\frac{2}{w_i}g^2(x_i; \tau^*)] \times (z - G(x; \tau))$$

$$(4.27)$$

where τ^* is a known value traced from the previous iteration. In stage 3 of the CME (see Algorithm 4.1), we replace the estimate of τ by reweighting the generalized least squares estimate of 4.27. Therefore the final classical least squares multistage estimate (CLsME) computation procedure can be performed using the following algorithm.

Algorithm 4.3 Classic least squares multistage estimate.

Stage 1: Compute the least squares estimate of θ.

Stage 2: Use the parameter estimate of θ in stage 1 to compute the sample variances z_i using Equation 4.18.

Stage 3: Use the parameter estimate of θ in stage 1 and sample variances computed in stage 2 to calculate the reweighing generalized least squares estimate $\hat{\tau}_{OLS}$ by Equation 4.27.

Stage 4.1: Estimate the variance function H ($\hat{\sigma}_i^2 = G(x_i; \hat{\tau})$) in Equation 4.8 using the estimate of $\hat{\tau}$ obtained in stage 3.

Stage 4.2: Calculate the estimate of the covariance matrix of ε in Equation 4.13 (denoted by $diag[G(x_i; \hat{\tau})] = \hat{\sigma}^2 \hat{V}$).

Stage 4.3: Estimate the parameter q ($\hat{\theta}_{CLsME}$) by the feasible generalized least squares (Section 2.3.3) using the estimate of the covariance matrix $\hat{\sigma}^2 \hat{V}$ in stage 4.2.

4.7 Robust Least Squares Estimate of the Structural Variance Parameter

In an attempt to implement a robust form of Algorithm 4.3, everything is kept the same as Algorithm 4.2 except the structural variance parameter τ estimate in stage 3. In this stage, a new robust form of the reweighting generalized least squares step of Equation 4.27 is produced by replacing it with a generalized M-estimate (3.21). The MM-estimate for τ and σ requires finding the minimum of the loss function given by

$$\hat{\tau}_M = \arg \min_{\tau} \sum_{i=1}^{n} \rho \left[\frac{(z_i - H(x_i; \tau))}{\hat{\sigma}_M^2 \hat{\gamma}(x_i; \tau^*)} \right]$$

$$(4.28)$$

where $\hat{\gamma}(x_i; \tau^*) = \sqrt{2/w_i}g(x_i; \tau^*)$ is the weight that is updated during the two iterations of Equation 4.28, and τ^* is a known value traced from a previous iteration. The generalized robust multistage algorithm form of 4.3 can be expressed as shown in Algorithm 4.4.

Algorithm 4.4 Robust generalized multistage estimate.

Stage 1: Compute the MM-estimate of θ by Algorithm 3.1.
Stage 2: Use the stage 1 estimate to compute the robust sample variances z_i using Equation 4.21.
Stage 3: Use the parameter estimate of θ in stage 1 and sample variances computed in stage 2 to estimate the parameter τ using the robustified MM-estimate in Equation 4.28.
Stage 4.1: Estimate the variance function G $(\hat{\sigma}_i^2 = G(x_i; \hat{\tau}))$ in Equation 4.13 using the estimate of $\hat{\tau}$ obtained in stage 3.
Stage 4.2: Calculate the estimate of the covariance matrix of ϵ in Equation 4.13.
Stage 4.3: Estimate the parameter θ $(\hat{\theta}_{RGME})$ by reweighing the generalized MM-estimator of Equation 4.28.

4.8 Weighted M-estimate

Lim et al. (2010) proposed the weighted M-estimate (WME), which involves simultaneously minimizing the robustified form of the likelihood estimate:

$$\begin{pmatrix} \hat{\theta} \\ \hat{\tau} \end{pmatrix} = \arg \min_{(\theta,\tau)} \sum_{i=1}^{n} \left\{ \rho \left(\frac{y_i - f(x_i; \theta)}{\sigma(x_i; \tau)} \right) + \log \sigma(x_i; \tau) \right\} : \theta \in R^p, \tau \in R^q$$

$$(4.29)$$

This estimate is a generalized form of the maximum likelihood estimate (4.15), in which the squares of errors are replaced by the downgrading loss function ρ. Lim et al. (2012) showed that this estimate is asymptotically normal.

We display all variables and gradients as vectors and matrices. Let $\epsilon = (\mathbf{Y} - \mathbf{f}(x; \theta))/\mathbf{G}(x; \tau)$. Then $E(\psi(\epsilon)\epsilon) = \gamma_1(\neq 0)$, $E(\psi'(\epsilon)) = \gamma_2(\neq 0)$, $E(\psi'(\epsilon)\epsilon^2) = \gamma_3(\neq 0)$, and $E(\psi^2(\epsilon)) = \sigma_{\psi_1}^2 < \infty$. Define the following values:

$$\kappa(x_i; \tau) = \frac{1}{G(x_i, \tau)}$$

$$\Gamma_{1n}(\hat{\theta}, \hat{\tau}) = \gamma_2 \sum_{i=1}^{n} \kappa^2(x_i; \tau)\hat{\mathbf{f}}(x_i; \theta)\dot{\mathbf{f}}^T(x_i; \theta)$$

$$\Gamma_{2n}(\hat{\theta}, \hat{\tau}) = \sum_{i=1}^{n} \left\{ \frac{2\gamma_1 + \gamma_3 - 1}{G(x_i; \tau)} \dot{G}(x_i; \tau)\dot{G}^T(x_i; \tau) + \frac{1 - \gamma_1}{G(x_i; \tau)} \ddot{G}(x_i; \tau) \right\}$$

$$\Gamma_{31n}(\hat{\theta}, \hat{\tau}) = \sigma_2 \sum_{i=1}^{n} \kappa^2(x_i; \tau)\dot{\mathbf{f}}(x_i; \theta)\dot{\mathbf{f}}^T(x_i; \theta)$$

$$\Gamma_{32n}(\hat{\theta}, \hat{\tau}) = \sigma_{\psi^2}^2 \sum_{i=1}^{n} \kappa^2(x_i; \tau)\dot{G}(x_i; \theta)\dot{G}^T(x_i; \theta)$$

where Γ_{31n} and Γ_{32n} are positive definite matrices. The asymptotic distribution of the weighted M-estimate is:

$$\hat{\Gamma}^{-\frac{1}{2}}\sqrt{n}\left(\begin{array}{c} \hat{\theta} - \theta \\ \hat{\tau} - \tau - \nu_n(\theta, \tau) \end{array} \right) \to \mathbf{N}_{p+q}(\mathbf{0}, \mathbf{I}_{p+q})$$

where

$$\nu_n(\theta, \tau) = \left(\frac{1}{n}\hat{\Gamma}_{2n}(\hat{\theta}, \hat{\tau})\right)^{-1}\frac{\gamma_1 - 1}{n}\sum_{i=1}^{n}\kappa(x_i; \tau)\dot{G}$$

$$\hat{\Gamma} = \left(\frac{1}{n}\hat{\Gamma}_{5n}(\hat{\theta}, \hat{\tau})\right)^{-1}\left(\frac{1}{n}\hat{\Gamma}_{3n}(\hat{\theta}, \hat{\tau})\right)^{-1}\left(\frac{1}{n}\hat{\Gamma}_{5n}(\hat{\theta}, \hat{\tau})\right)^{-1}$$

$$\hat{\Gamma}_{3n}(\theta, \tau)\left[\begin{array}{cc} \Gamma_{31n}(\theta, \tau) & 0 \\ 0 & \Gamma_{32n}(\theta, \tau) \end{array} \right]$$

and

$$\hat{\Gamma}_{5n}(\theta, \tau)\left[\begin{array}{cc} \Gamma_{1n}(\theta, \tau) & 0 \\ 0 & \Gamma_{2n}(\theta, \tau) \end{array} \right]$$

Argument variables are omitted for simplicity. Note that the above argument provides the asymptotic covariance matrix of the WME. This will be used for computational purposes in the `nlr` package.

4.9 Chicken-growth Data Example

Appendix A explains the detail of the data collection method and sampling errors for the chicken-growth data presented by Riazoshams and Miri (2005). Table A.1 displays the data. As shown in Figure 4.1, the data variance increases over time. This is a natural phenomenon that occurs due to the increase in weight during the chickens' lifetime, which logically causes the variation of the data to increase. The heteroscedasticity of the data due to this natural behavior is therefore obvious and should be taken into consideration when modeling it.

The growth rate seen in the chicken data is S-shaped: for young chickens, weight increases slowly, then it rapidly increases at adolescence, before declining on maturity. Finally, the weight tends to a constant value at infinity. This behavior creates an S-shaped growth curve.

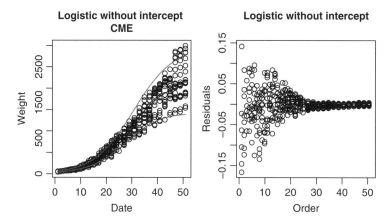

Figure 4.1 Chicken-growth data estimates using the classical multistage estimator.

In practice, many models can be fitted to the data. Practical considerations and statistical value criteria can be used to select a good one. Experts in the subject problem may use physical techniques to identify a suitable nonlinear model, or differential equations could be used. Statistically, a model with a smaller number of parameters, sum of errors, and better selection model criteria (such as the cross-validation (see Bunke et al. (1995a)), Bayesian selection criterion, and Akaike information criterion) is preferred. For the chicken-growth data the logistic model

$$y_i = \frac{a}{1 + be^{-cx_i}} \tag{4.30}$$

was chosen by Riazoshams and Miri (2005).

To fit the model properly to the heteroscedastic variance, weighted least squares, variance modeling, or transformation of one or both sides of the model might be used. To find the right way, the empirical variance of the data can be helpful. At each date i, the empirical variance can be computed by:

$$s_i^2 = \frac{1}{n_i - 1} \sum_{j=1}^{n_i} (Y_{ij} - \overline{Y}_i)^2. \tag{4.31}$$

A standardized form of errors is defined as

$$e_{ij} = \frac{Y_{ij} - f(\mathbf{X}_i; \hat{\theta})}{s_i}$$

The weighted least square estimator minimizes the weighted sum of squares, such that a power of variances is considered as the weight:

$$s(\theta) = \sum_{i=1}^{n} \sum_{j=1}^{n_i} \frac{(Y_{ij} - f(\mathbf{X}_i; \hat{\theta}))^2}{(\sigma_i^2)^\tau}$$

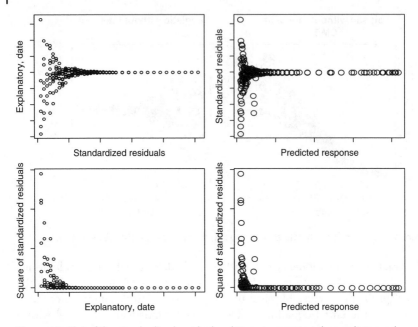

Figure 4.2 Plot of the standardized residual and its square against the predictor and response to assess the adequacy of the weighted least squares of the model. *Source*: Riazoshams and Midi (2009). Reproduced with permission of European Journal of Science.

To find an adequate form for the weight, the standard values e_{ij} must be independent random variables with mean 0 and variance 1. In addition, in order to find an appropriate power, we plot standardized residuals and their powers versus response and independent variables as well.

Figure 4.2 shows the plot of the standardized residuals and their power with respect to the explanatory and response variables. It can be seen that the standardized residuals are not randomly scattered. This suggests that the weighted least squares approach is not appropriate.

Since the weighted least squares is not appropriate we can check for any possibility of variance modeling. Six variance models discussed by Bunke et al. (1995a) are shown below.

Model 1: Power function
$$\sigma_i^2 = \sigma^2 f^\lambda(x_i; \theta) \ or$$
$$(= \sigma^2(f^\lambda(x_i; \theta) + a)) \tag{4.32a}$$

Model 2: Exponential
$$\sigma_i^2 = \sigma^2 \exp(\lambda f(x_i; \theta)) \tag{4.32b}$$

Model 3: Linear
$$\sigma_i^2 = \sigma 2(1 + \lambda f(x_i; \theta)) \tag{4.32c}$$

Model 4: Unimodal quadratic

$$\sigma_i^2 = \sigma^2 + \tau_1(f(x_i; \hat{\theta}) - \overline{f})$$
$$+ \tau_2(f(x_i; \hat{\theta}) - \overline{f})^2$$
$$\overline{f} = \frac{1}{n} \sum_{i=1}^{n} f(x_i; \hat{\theta})$$

(4.32d)

Model 5: Bell-shaped variance

$$\sigma_i^2 = \sigma^2 + \sigma_\tau^2(\overline{f} - f_{m_0}(x_i; \tilde{\theta}_{m_0})$$
$$\times (f_{m_0}(x_i; \tilde{\theta}_{m_0}))$$

(4.32e)

Model 6: Simple linear $\qquad \sigma_i^2 = \sigma 2 + \tau(f(x_i; \theta) - \overline{f})$ (4.32f)

There are several ways to choose an appropriate model. If the data are replicated, as in the chicken-growth data example, the empirical variances can be a primary estimate for variances σ_i^2 and, using the properties of the above models, graphical methods can be used to find an appropriate model. If this method does not work, there are a few selection model criteria that can be used. Bunke et al. (1995a, 1999) discussed the cross-validation selection criterion and provided some SPLUS tools for computational purposes (see also Bunke et al. (1995b, 1998)). To identify an appropriate model using graphical techniques, empirical variances s_i^2 (4.31) can be used as an estimate of variance σ_i^2, and response averages $\overline{Y}_i$ can be used as estimate of function model $f(x_i; \theta)$.

For the power function model (4.32a), we have $\log(\sigma_i^2) \propto \lambda \log(f(x_i; \theta))$ so the logarithm of the computed empirical variance is in proportion to the logarithm of the response average, which means that their graphs against each other exhibit a linear regression.

For the exponential model (4.32b), we have $\log(\sigma_i^2) \propto f(x_i; \theta)$ so the logarithm of the computed empirical variance is in proportion to the response average, which means that their graphs against each other exhibit a linear regression.

For the linear and simple linear models (4.32c and 4.32f), the computed empirical variance is in proportion to the response average, so the graph of empirical variance against response average exhibits a linear regression.

For the unimodal and bell-shaped models (4.32d and 4.32e), the computed empirical variance is quadratically related to the response average, which means that the graph of empirical variance against response average exhibits a quadratic regression relationship.

For the chicken-growth data, since the data are replicated, the empirical variances and response averages can be computed easily, and the inferences above can be performed. Figure 4.3 shows several such graphs. Figure 4.3b shows that

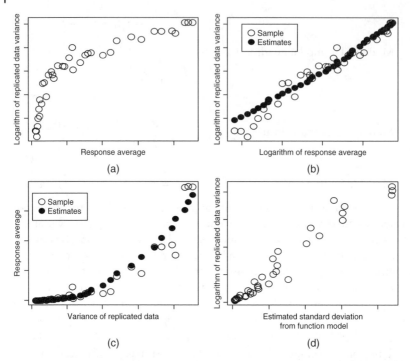

Figure 4.3 Graphs for choosing the model variance: (a) exponential variance model assessment; (b) power variance model assessment; (c) adequacy of estimated variance model; (d) assessment of the equality of standard deviation and estimate thereof.

the logarithm of sample variance, computed at each date, is linear with respect to the logarithm of the response average, which shows that the variance model can be a power function of the nonlinear model (4.32a). Figure 4.3a, which is the graph of the variance logarithm against the response mean, is not linear, so the exponential model (4.32b) is inappropriate. Figure 4.3d shows a plot of the estimated standard deviation from a fit of the logistic nonlinear regression model against the empirical standard deviation, which reveals that the variance power model can be truly fitted to the data. Therefore the logistic model with variance as a power function of the nonlinear model can be considered:

$$
\sigma_i^2 = \sigma^2 f^\tau(x_i; \lambda)
$$

$$
= \sigma^2 \left[\frac{a}{1 + be^{-cx_i}} \right]^\lambda \tag{4.33}
$$

Since the data do not include outliers, the parameters are estimated using the CME. The estimated values and standard errors of parameters are shown in Table 4.1. For comparison purpose, the robust estimates using the robust

Table 4.1 Parameter estimates for chicken-growth data using a logistic model with a power heteroscedastic function.

Parameters	Estimates	Standard errors	95% confidence interval
CMS			
a	2209.836	2608.6	(2062.45, 2357.22)
b	51.37095	1.1873	(47.70, 55.04)
c	0.1327036	0.10963	(0.13, 0.14)
λ	2.340179		
σ^2	1.040946		
RMS			
a	2245.382	1.1663	(2241.78, 2248.99)
b	51.87051	0.03121	(51.77, 51.97)
c	0.1315638	4.111e-05	(0.13, 0.13)
λ	2.092687		
σ^2	1.816597		

multistage method are also presented in this table. The fitted graph with confidence bounds and residual plots is shown in Figure 4.1. Note that, due to heteroscedasticity, the residuals after transformation by the heteroscedastic variance should be random, not the original residuals shown in the figures.

4.10 Toxicology Data Example

Lim et al. (2010, 2012) illustrated the WME methodology with real data from a national toxicology study program (Bucher 2007). The Hill model, defined by

$$y_{ij} = \theta_0 + \frac{\theta_1 x_i^{\theta_2}}{\theta_3^{\theta_2} + x_i^{\theta_2}} + \sigma(x_i; \tau)\varepsilon_{ij}, \ i = 1, \dots, 7, j = 1, \dots, 4$$

was used to fit the relationship between dose concentration x and chromium concentration y in the blood and kidneys of a mouse data set collected by Bucher 2007. Lim et al. (2010, 2012) showed that the heteroscedastic variance can be expressed in a nonlinear form:

$$\sigma(x_i; \tau) = \tau_0 + \frac{\tau_1}{1 + e^{-\tau_2 x_i}}, i = 1, \dots, n$$

Table 4.2 Parameter estimates of the Hill model for chromium concentration in the kidney of mouse (values in parentheses show estimated errors).

Parameters	CME	RME	CLsME	RGME	WME Square function	WME Hampel function
θ_0	0.200	0.0987	0.120	0.098	0.103	0.152
	(0.072)	(0.031)	(0.041)	(0.074)	(0.094)	(0.017)
θ_1	3.765	2.605	3.509	2.995	4.105	3.753
	(0.523)	(0.795)	(0.818)	(4.639)	(0.731)	(0.699)
θ_2	1.722	0.980	1.068	0.961	1.133	1.220
	(0.528)	(0.179)	(0.165)	(0.640)	(0.400)	(0.201)
θ_3	37.450	28.679	39.653	31.840	45.835	48.927
	(10.440)	(18.257)	(17.894)	(105.258)	(18.753)	(20.373)
τ_0	−0.673	−0.734	−1.475	−0.883	−0.956	−0.748
τ_1	1.967	1.397	2.747	1.733	2.032	1.557
τ_2	0.037	0.038	0.0334	0.049	0.036	0.037
σ	0.685	12.541	0.969	3.553	0.754	0.811

Since the gradient of the function model, with respect to θ_2 at point $x_i = 0$, does not exist, the Nelder–Mead simplex method is used to optimize the problem in this example.

The CME, RME, CLsME, RGME, and WME are shown in Table 4.2, and the fitted models are shown in Figure 4.4. The WME is calculated once using the

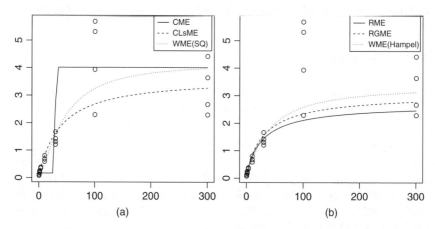

Figure 4.4 Toxicology data, fitted by robust and classical methods: (a) classical estimates, (b) robust estimates. *Source*: The Performance of a Robust Multistage Estimator in Nonlinear Regression with Heteroscedastic, Errors, Communications in Statistics – Simulation and Computation 2016. Adapted from Simulation and Computation 2016.

square function ρ, which is not robust, and once using the Hampel function, which is an example of a robust method. The computations in this chapter can be performed with the `nlr` package, as explained in Chapter 8.

4.11 Evaluation and Comparison of Methods

Outliers in nonlinear regressions of data sets with heteroscedastic variance have unexpected effects. When the variance at a cross section of a predictor is much larger or smaller than the variance function model value, a new concept is encountered: the variance outlier. The data in a nonlinear regression curve might be spread over a cross-section of response values and cause the large variance. Just two data points is enough for such an effect, but the situation will be more complicated if, at a cross-section, the number of outlier points is more than 50% of the number of replicate data points and therefore we cannot identify which data is good and which is bad. In this case, outlier data and variance outliers occur at the same time, but we cannot determine which data are the outliers. The strangest behavior occurs when the data in one cross-section of the predictor values are not outliers, but replicated data squeezed at that point and resulting in too small a variance (see Figure 6.2). In this case the data points might be due to an abnormal experiment, but are not outliers, while their variances are outliers, although extremely small. In all of these cases, the variances are outliers and will lead us to a new term: variance outlier.

Riazoshams and Midi (2016) were the first to introduce variance outlier definitions and they discussed the differences between several methods. Different kinds of outliers and their detection are discussed in Section 6.7. In this section the methods are compared.

The convergence distribution of the WME was proved by Lim et al. (2010, 2012), who proposed the WME method. It can be used to fit models with multivariate predictor variables and nonreplicated data. Riazoshams and Midi (2016), who proposed the RME and RGME, compared these models by simulation. Since simulation methods are limited, they cannot be extended to all cases until the mathematical behavior of the methods has been studied. Computationally, however, RGME and RME are simpler. Riazoshams and Midi (2016) showed that, computationally, RME is more robust and efficient in some cases because it is more resistant to the effect of outliers. In the situations where predictors are multivariate, not replicated, or when computation is possible, the WME is preferred, but for complicated nonlinear regression models or where outliers are critical, RME is slightly more robust. RGME has a robustness measure close to that of RME. Although WME is more difficult to use because it optimizes simultaneously, using good optimization methods requires fewer iteration stages and, numerically speaking, it has fewer rounding errors and transfers of errors between stages.

Since, in nonlinear regression, parameter estimate iterations require initial values, and for complicated models achieving convergence is difficult, researchers might have to use all of the methods and, after comparison of the convergence results, choose the best one. Along with method differences, convergence tolerances, iteration limits, and initial values are essential in practice. All these factors will be involved in helping the analyst to choose the preferred method for computation.

5

Autocorrelated Errors

In Chapter 4 we discussed a null condition for which the regularity assumption – of homogeneity of error variances – does not hold. Another situation that violates the classical assumption is when the errors are autocorrelated. This situation might occur when the data are collected over time. In such a case it might be expected that time series methods could be applied. In fact, when the data over time follow a structural curve, we use nonlinear regression. However, the nonlinear structure of data are considered as a trend in the time series approach and the trend is removed by filtering. In this chapter we assume the errors of a nonlinear regression model are correlated, and attempt to find estimates using classical and robust methods. The theory, and also the computational tools provided in the `nlr` package, are based on the methods developed by Riazoshams et al. (2010).

5.1 Introduction

Some statistics practitioners often ignore the underlying assumptions when analyzing real data and employ the ordinary least squares (OLS) method to estimate the parameters of a nonlinear model. In order to make reliable inferences about the parameters of a model, the underlying assumptions, especially the assumption that the errors are independent, must be satisfied. However, in real situations we may encounter dependent error terms, which might produce autocorrelated errors.

A two-stage estimator (known as the classical two-stage, or CTS estimator) has been developed to remedy this problem. However, it is now evident that outliers have a significant effect on least squares estimates, so we expect that the CTS will also be affected by outliers since it is based on the least squares estimator, which is not robust. Riazoshams et al. (2010) proposed a robust two-stage (RTS) procedure for the estimation of the nonlinear regression parameters in data sets affected by both autocorrelated errors and outliers. Their numerical

Robust Nonlinear Regression: with Applications using R, First Edition.
Hossein Riazoshams, Habshah Midi, and Gebrenegus Ghilagaber.
© 2019 John Wiley & Sons Ltd. Published 2019 by John Wiley & Sons Ltd.
Companion website: www.wiley.com/go/riazoshams/robustnonlinearregression

example and simulation study show that the RTS method is more efficient than the OLS and CTS methods.

Consider the general nonlinear model defined in Chapter 2 and recall Equation 2.2:

$$y = \eta(\theta) + \varepsilon \tag{5.1}$$

This follows the same notation as used in Chapter 2. In classical cases it is assumed that the errors are identically independent distributed (i.i.d.) with mean zero and constant variance; that is, $E(\varepsilon) = 0$, $Var(\varepsilon) = \sigma^2 I$. A common problem is known as an autocorrelated error, and it occurs when each error is correlated with previous errors. This problem usually happens in the situation when the data are collected over time (see Grassia and De Boer (1980) and White and Brisbin Jr (1980)). Unfortunately, many statistics practitioners are not aware that analyzing such data using OLS has many drawbacks. Seber and Wild (2003) discussed the CTS method to rectify this problem.

The problem is further complicated when the violation of the independent error terms occurs alongside the existence of outliers. Outlier(s) form a single or group of observations that are markedly different from the bulk of the data or from the pattern set by the majority of the observations. The OLS technique ignores the outliers and erroneously assumes that the errors are independent. Consequently, fewer efficient estimates are obtained. However, the CTS method alone cannot rectify the problem of both outliers and autocorrelated errors.

This problem prompted the development of a new and more efficient estimator that can rectify these two issues simultaneously. Sinha et al. (2003) proposed the generalized M (GM) estimator to estimate the parameters of the model when the errors follow an autoregressive (AR) process. Riazoshams et al. (2010) proposed an alternative method that incorporates the CTS method and the highly efficient and high breakdown point MM-estimator. It can to estimate the parameters of a nonlinear model with errors can be from any time series process; not just an AR process. They call this modified technique the RTS estimator. The RTS method is more efficient than the OLS and CTS methods since it removes the influence of outliers and deals with autocorrelation problems by using the highly robust MM-estimator in the CTS procedure.

5.2 Nonlinear Autocorrelated Model

When the errors are correlated, model 5.1 can be written as

$$y_i = f(x_i; \theta) + \varepsilon_i, \ i = 1, ..., n$$

where the ε_i are correlated and stationary, and the autocovariance and autocorrelation functions are defined as

$$\gamma_k = Cov(\varepsilon_i, \varepsilon_{i\pm k}), \ k = 0, 1, \dots$$

$$\rho_k = Corr(\varepsilon_i, \varepsilon_{i\pm k}) = \gamma_k/\sigma^2, \ k = 0, 1, \dots$$

respectively. Then the correlation matrix of ε_i, denoted by V, is given by

$$V = \begin{bmatrix} \rho_0 & \rho_1 & \cdots & \rho_{n-1} \\ \rho_1 & \rho_0 & \cdots & \rho_{n-2} \\ \vdots & \vdots & \vdots & \vdots \\ \rho_{n-1} & \rho_{n-2} & \cdots & \rho_0 \end{bmatrix} \tag{5.2}$$

As an example, assume the autocorrelation form follows the AR process, which is defined as

$$\varepsilon_i = \sum_{r=0}^{\infty} \psi_r a_{i-r}, \ \psi_0 = 1$$

The a_i are i.i.d. with mean 0 and variance σ_a^2. It is assumed that $\sum_{r=0}^{\infty} \psi_r < \infty$. The autoregressive error of degree (q) is defined as;

$$\varepsilon_i = \phi_1 \varepsilon_{i-1} + \phi_2 \varepsilon_{i-2} + \dots + \phi_q \varepsilon_{i-q} + a_i \tag{5.3}$$

with

$$\sigma^2 = \frac{\sigma_a^2}{1 - \phi_1 \rho_1 - \phi_2 \rho_2 - \dots - \phi_q \rho_q}$$

We note that $\gamma_0 = \sigma^2$ and $\rho_k = \gamma_k/\gamma_0$, so then the autocovariance and autocorrelation functions for AR(q) take the forms:

- $\gamma_k = \phi_1 \gamma_{k-1} + \phi_2 \gamma_{k-2} + \dots + \phi_q \gamma_{k-q} \ k>0$
- $\rho_k = \phi_1 \rho_{k-1} + \phi_2 \rho_{k-2} + \dots + \phi_q \rho_{k-q} \ k>0,$

respectively, when $\rho_0 = 1$. Therefore, the correlation matrix of ε_i is a function of parameter ϕ, (V_ϕ).

5.3   The Classic Two-stage Estimator

In the situation where autocorrelated errors exist, OLS is not an efficient estimator. It has been noted in the previous section that the CTS method was put forward to rectify the problem of autocorrelation. The CTS method is defined as follows:

Algorithm 5.1 Classic two-stage estimate.

Stage 1: (i) Fit model 5.1 by the ordinary nonlinear least squares method to obtain $\hat{\theta}_{OLS}$.

(ii) Compute the residuals $\hat{\varepsilon}_i = y_i - f(x_i; \hat{\theta}_{OLS})$.

(iii) Obtain an estimate $\hat{\phi}$ of ϕ from $\{\hat{\varepsilon}_i\}$ and hence calculate $V_{\hat{\phi}}^{-1}$.

Stage 2: Choose the two-stage estimator $\hat{\theta}_{TS}$ to minimize

$$[y - f(\theta)]^T V_{\hat{\phi}}^{-1} [y - f(\theta)]$$

The final estimates, denoted by $\hat{\theta}_{CTS}$, are obtained by employing the feasible generalized least squares procedure as described in Section 2.3.3.

5.4 Robust Two-stage Estimator

Seber and Wild (2003) described a two-stage estimation procedure to rectify the problem of autocorrelated errors. We refer to this procedure as the CTS procedure. The main limitation of this estimator is that it depends on the classical estimation technique to estimate the parameters of a model. Although the CTS method can rectify the autocorrelated errors, it is not robust when contamination occurs in the data.

The CTS procedure cannot handle both these problems at the same time. In this situation, we need to develop a method that can remove or reduce the effects of both outliers and autocorrelated errors. In this respect, Riazoshams et al. (2010) propose incorporating robust MM- and M-estimators in the formulation of the RTS procedure. The algorithm for the RTS estimator is similar to that for the CTS except that all the classical estimation techniques used to estimate the parameters in each stage are replaced by robust estimators. The RTS algorithm is computed as follows:

Algorithm 5.2 Robust two-stage estimate.

Stage 1: (i) Fit model 5.1 by using the MM-estimator to obtain $\hat{\theta}_{MM}$.

(ii) Compute the residuals $\hat{\varepsilon}_i = y_i - f(x_i; \hat{\theta}_{MM})$.

(iii) Obtain an estimate $\hat{\phi}$ of ϕ from $\{\hat{\varepsilon}_i\}$ using the robust M-estimator and hence calculate $V_{\hat{\phi}}^{-1}$ from (5.2).

Stage 2: The final estimates, denoted by $\hat{\theta}_{RTS}$, are obtained by employing the generalized MM-estimate, as in (3.21).

The generalized M-estimator is used for the estimation of the autoregressive process parameters (see Martin (1980) and Maronna et al. (2006)). The general form of the M-estimator for an ARMA(p,q) process, which includes the AR(p) process, is computed by minimizing

$$\sum_{t=p+1}^{n} \rho\left(\frac{\hat{u}_t}{\hat{\sigma}_a}\right)$$

where ρ is an influence function. As mentioned before, there are several possible choices for ρ functions (for example see Hoaglin et al. (1983) and Maronna et al. (2006)). $\hat{u}_t(\lambda)$, as described in Maronna et al. (2006), represents the residuals and for the AR(p) process it can be written as:

$$\hat{u}_t = \hat{u}_t(\phi) = \varepsilon_t - \phi_1 \varepsilon_{t-1} - \dots - \phi_1 \varepsilon_{t-q}$$

where $\hat{\sigma}_a$ is a robust scale estimate.

The CTS and RTS approaches assume the correlation structure of errors is known, but in practice it is not, so we have to identify the correlation structure. When the data follows a nonlinear regression in a time series this is called a trend. Time series methods suggest the trend should be removed and then the structure of the autocorrelation can be found. In the nonlinear regression approach in the presence of autocorrelation, the same thing can be done, but some problems may arise.

At the beginning of nonlinear regression modeling we have to find the nonlinear structure of the data. If the structure is known, we first have to estimate the parameters and fit the model, and then compute the residuals. Finally, the residuals can be used to find the autocorrelation structure of the errors, using the CTS or RTS methods. This might work there are no outliers, but if there are even using the robust method might not work. We have not found a single robust estimation method for all situations; instead, we have used a set of methods, including the robust estimate, deleting outliers and OLS estimation. The next section describes an example.

5.5   Economic Data

In this section we analyze the Iranian net money data for 1960–2010, which are shown in Table A.5 (see also Example 8.10), and the Iranian trademark application (direct resident) data from 1960–2006, shown in Table A.6. The data were obtained from the World Bank (data.worldbank.org; but note that data for before 1960 are not available directly from the homepage).

Since both of the data sets were collected over time, they might be correlated, but Example 8.10 shows that the net money data are uncorrelated. Therefore, in this chapter, we fit models to the trademark data. It is noteworthy that there are a few ways of analyzing these data. One approach is to use linear regression by taking logarithms of one or both of the variables. Alternatively, the data can be considered as a time series with an exponential trend, which is similar to our approach in this chapter. Because of the existence of outliers, this chapter attempts to fit a nonlinear regression model using robust methods with autocorrelated errors.

There are three steps that should be followed to find a nonlinear model for the data, as shown in Algorithm 5.3.

Algorithm 5.3 Fitting a nonlinear regression model with autocorrelated errors.

Stage 1: Fit a nonlinear regression model to the data. Since the data include outliers, we can apply robust methods or delete the outliers. Deleting data in a nonlinear regression is not correct in all cases because it changes the shape of the model curve.

Stage 2: Perform a residual analysis. Compute the residuals and try to identify the autocorrelation structure. Note that using the robust method reduces the sensitivity of the autocorrelation structure and results in large residuals for outliers. Thus, a residual analysis using a nonrobust method (by deleting outliers), or performing residual analysis by robust methods, might ensure a correct result. As mentioned earlier, such an approach depends on the position of the data and the shape of the nonlinear model.

Stage 3: Use the identified correlation structure to fit the model using a two-stage method (CTS) or (RTS) for data with outliers.

First, a nonlinear model should be found for the data. For the Iran trademark data the programs using the nlr package are shown in Section 8.9. There are several possible nonlinear models that might fit the data. The inferences for the four models used for the methane and carbon dioxide data examples, and defined in Equation 3.22, are shown in Table 5.1. This shows that the fourth model, the exponential with intercept model, is preferred due to its lower number of parameters, lower correlation between parameters and lower loss function value. It is important to note that its higher number of iterations – 26 compared with 7 for the first model – might cause some convergence problems in the next stage. Researchers could therefore choose the first model instead if faced with non-convergent iterations. With the same inferences, the same model can be used to analyze the net money data as well.

Figure 5.1 shows the fit using both the least squares and robust MM methods. The curves of the two fits are similar because the outlier is surrounded by other points, which pull the curve toward them, even in the OLS estimate. However, it can be seen that the difference is larger for an outlier in the x-direction, among the last few points at the right-hand side. For example, see Figure 5.2 for the Iran net money data. In this case the outlier point is on its own and pulls the tail of the nonlinear curve towards it, which changes the shape of the curve more than in the first example. This reveals that the point position is important in

Table 5.1 The robust MM-estimates of fitted four models for the Iran trademark data.

Scaled exponential		Achieved loss function: 35.7505				
Parameters	MM	Standard error	Correlation matrix			
p_1	1351.010	1.034070×10^{-3}	1.	−0.088	0.088	−0.554
p_2	22604.47	2.217505×10^{-10}		1	−0.999	−0.481
p_3	2005.987	9.054273×10^{-7}			1	0.481
p_4	5.536163	1.676315×10^{-6}				1
σ	559.4517					
Correlation	0.9481255					
Iterations	7					
Scaled exponential convex		Achieved loss function: 35.72642				
Parameters	MM	Standard error	Correlation matrix			
p_1	1351	3.725483×10^{-6}	1	0.555	0.554	
p_2	352.3076	3.949508×10^{-7}		1	0.999	
p_3	0.1806261	1.970270×10^{-10}			1	
σ	559.4614					
Correlation	0.9481249					
Iterations	9					
Power model		Achieved loss function: 42.97959				
Parameters	MM	Standard error	Correlation matrix			
p_1	0.002392549	1.101957×10^{-8}	1	0.000	0.000	
p_2	$-4.535917 \times 10^{-70}$	4.907359×10^{-74}		1	0.999	
p_3	1.08795	5.898310×10^{-8}			1	
σ	1122.364					
Correlation	0.4579801					
Iterations	8001 (tolerance not achieved)					
Exponential with intercept		Achieved loss function: 35.72252				
Parameters	MM	Standard error	Correlation matrix			
p_1	1350.998	6.194030×10^{-6}	1	0.573	−0.554	
p_2	1950.48	9.818122×10^{-8}		1	−0.999	
p_3	5.536321	1.004122×10^{-8}			1	
σ	559.4629					
Correlation	0.9481248					
Iterations	26					

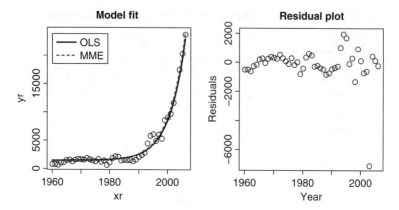

Figure 5.1 Fitted exponential with intercept model for the Iran trademark data using OLS and MM-estimates.

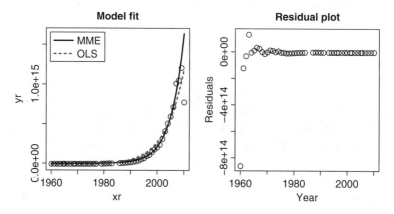

Figure 5.2 Fitted exponential with intercept model for the Iran net money data using OLS and MM-estimates.

determining how much the the shape of the curve changes. In contrast, in linear regression the outlier only rotates or moves the line but does not change the shape. Point position is therefore an extra issue to be taken into consideration for nonlinear regression. We have dealt with this effect elsewhere in the book.

In the second step, the autocorrelation structure of the residuals has to be found and the parameters estimated. The residuals should be computed first from the fitted model, then the autocorrelation function (ACF) and the partial autocorrelation function (PACF) can be computed. These can then be used to identify the correlation structure. Residuals can be computed from both the MM-estimate and OLS. The MM-estimate will display a larger residual value for the outlying point. In both examples, there is only a single outlier, and the effect of the outlier is reduced by the surrounding points, although in the net money data it is higher and the residual of the outlying point computed from the OLS estimate is also large. In computing the ACF, the PACF will therefore be affected by outliers and result in incorrect inference. We therefore have to solve the problem of large values of outlier residuals. It is obvious that the residuals from the MM-estimate are more accurate and should be used for inference.

Four approaches are possible for dealing with the residual outlier. First, the outlier residual can be deleted and the ACF or PACF can then be calculated. This case is correct when the point position, as in the above two examples, is not critical.

The second approach is to delete the outliers, fit the model using MM or OLS, then compute the ACF and PACF. In this case the positions of the points are important.

The third approach is to use robust ACF and PACF methods. There are a few choices but R lacks tools for robust computation of ACF or PACF. Dürre et al. (2015) compared several robust methods. Shevlyakov et al. (2013) considered various robust modifications of the classical methods of power spectra estimation, both nonparametric and parametric. Smirnov and Shevlyakov (2010) proposed a parametric family of M-estimators of scale based on the Rousseeuw–Croux Qn-estimator, which is the basis for computation in the robcor R package, as discussed in Chapter 8.

The fourth solution is to predict the response value for the outlying data, obviously from a robust fit, then swap it for the outlier, and again fit the model with a new value. The residuals and consequently the ACF and PACF can then be calculated. The advantage of this method is that there will be no deleted index in the data that might create lag difference problems, but the outliers have to be identified first; for example the methods discussed in Chapter 6 can be used in critical situations. Comparing the results of all these approaches will lead us to the correct model.

Consider the Iran trademark data. Figure 5.3 displays the ACF, PACF, and robust ACF for MM- and OLS estimates. As can be seen from the ACF and PACF (Figure 5.3a–d), no special autocorrelation structure can be identified, which does not seem to be true from the residual plots in Figure 5.1. This means that the large values of the residuals for the outlier points for both estimators swamped the autocorrelation structure. Figure 5.3e,f shows the computed robust ACF. It pretends the residuals follow an ARIMA model, of order 2 for MA and 8 for seasonality.

Figure 5.4 displays the ACF, PACF, and robust ACF for MM- and OLS estimates when the residual of the outlier point is deleted. The ACF and PACF for the MM-estimate (Figure 5.4a,b) reveals an ARIMA(1, 0, 1)(0, 0, 1)7 model, which is nonseasonal autoregressive of order 1, moving average of order 1, and seasonal moving average with order 1 and seasonality 7. The ACF and PACF for the OLS estimate (Figure 5.4c,d) show the residuals as being swamped by the outlier, which is correct since the OLS is affected by the outlier from the

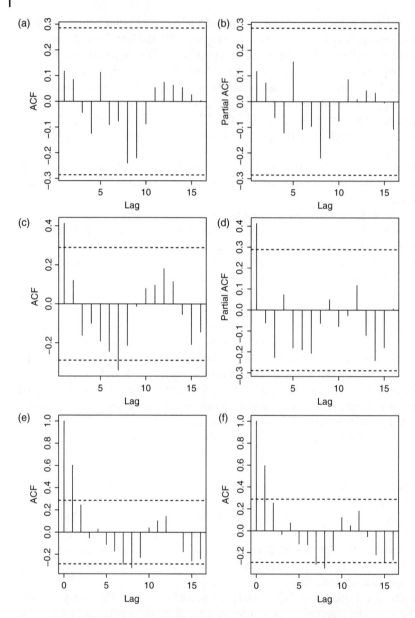

Figure 5.3 ACF, PACF, and robust ACF for OLS and MM-estimates. (a) MM-estimate residuals, (b) MM-estimate residuals, (c) OLS-estimate residuals, (d) OLS-estimate residuals, (e) MM-estimate residuals, (f) OLS-estimate residuals.

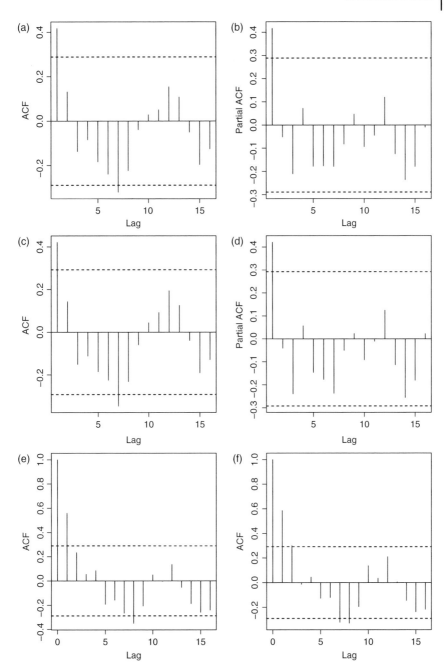

Figure 5.4 ACF, PACF, and robust ACF for OLS and MM-estimates. The residual of the outlier was deleted from the computed residuals from model fits. (a) MM-estimate, (b) MM-estimate, (c) OLS-estimate, (d) OLS-estimate, (e) robust ACF MM-estimate, (f) robust ACF OLS estimate.

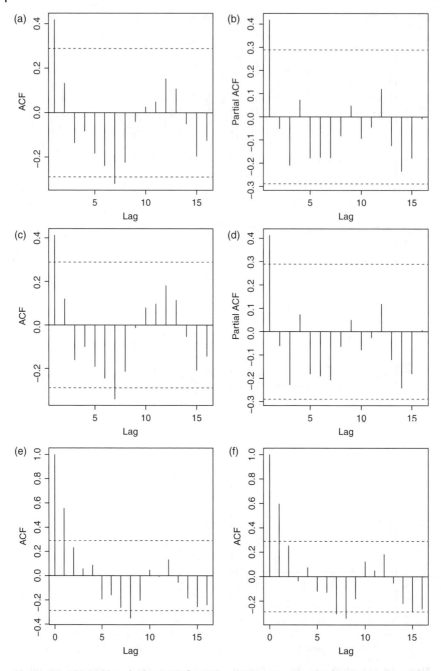

Figure 5.5 ACF, PACF, and robust ACF for OLS and MM-estimates. The fits were performed after the outlier had been deleted and the residuals were computed from the fitted model without the outlying point: (a) MM-estimate, (b) MM-estimate, (c) OLS-estimate, (d) OLS-estimate, (e) robust ACF, (f) robust ACF.

original fit. Figure 5.4e,f shows a robust ACF, from which it appears that the residuals follow an ARIMA model of order 2 for MA and 8 for seasonality, as before.

Figure 5.5 displays the ACF, PACF, and robust ACF for the MM- and OLS estimates for when the outlier is deleted from the data and the model is fitted to the cleaned data. Figure 5.6 displays the estimated models for clean data. The same inference as above can be made from this figure, and the same result is obtained for the OLS estimate because the outlier point has been removed.

Figure 5.7 displays the ACF, PACF, and robust ACF for the MM- and OLS estimate for when the outlier is predicted and the model fitted for all the data along with the predicted response. The same inference as above can be concluded from this figure. Moreover, the same model (ARIMA(1,0,1(0,0,1)7)) can be obtained from the OLS estimate because the outlier point is predicted and its effect removed. Again, the robust ACF (Figure 5.7e) displays the ARIMA model with moving average order 2 and seasonality 7. To select a better model, the Akaike information criterion for ARIMA(1,0,1)(0,0,1)7 of 730.22 is lower than for ARIMA(1,0,2)(0,0,8)7 at 739.23. The robust PACF is not provided in R-tools, so the ARIMA(1,0,1)(0,0,1)7 model, with seasonality 7 is chosen for the residuals of the data. The Iran trademark data follows the exponential with intercept model (model 4) with errors following an ARIMA(1,0,1)(0,0,1)7 process.

Note that the derived model for the Iran trademark data, ARIMA(1,0,1) (1,0,1)7, excludes any trend, indicating that in nonlinear regression with autocorrelated errors, the nonlinear regression model is the removed trend and the autocorrelated error is a seasonal time series. Such a model in a time series approach, for time variable z_t, is traditionally written as:

$$z_t = P_t + S_t + e_t$$

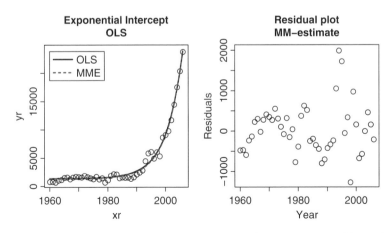

Figure 5.6 Model fit using OLS and MM-estimators when the response of the outlier point was replaced by its prediction.

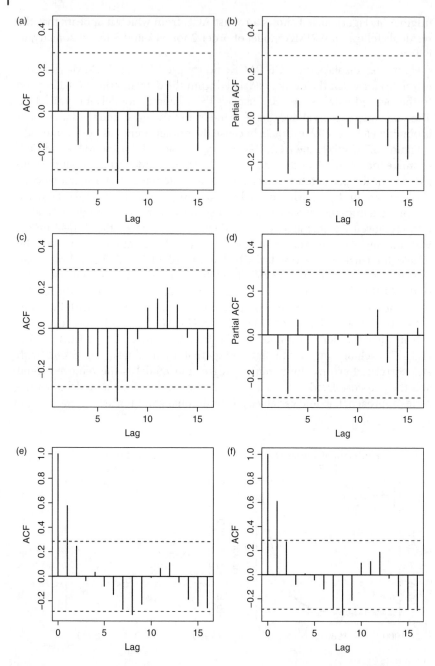

Figure 5.7 ACF, PACF, and robust ACF for OLS and MM-estimates. Residuals were computed from the fitted model when the response of the outlier point was replaced by its prediction. (a) MM-estimate, (b) MM-estimate, (c) OLS-estimate, (d) OLS-estimate, (e) robust ACF, (f) robust ACF.

Table 5.2 The RTS estimates of the fitted exponential with intercept models for the Iran trademark data, with ARIMA(1,0,1)(0,0,1)7 autocorrelated structure.

Scaled exponential	Achieved loss function: 46.9146				
Parameters	MM	Standard error	Correlation matrix		
p_1	1261.879	2.72196876	1.	0.598	−0.579
p_2	1946.466	0.04442580		1	−0.998
p_3	5.936317	0.00453923			1
ϕ	0.3803	0.2495			
θ	−0.0009	0.2498			
Θ	−0.2615	0.1443			
σ	548.983				
Correlation	0.9450576				
Iterations	7				

where P_t is the trend-cycle component (which in nonlinear regression is the nonlinear function model), S_t is the seasonal component, and e_t is an irregular component.

In the next step, the RTS method can be used to get the final fit. For this purpose we need to derive the analytical autocorrelation function for ARIMA(1,0,1)(1,0,1)7. Section 5.6 derives the formulas for the autocorrelation function for this seasonal ARIMA model. The parameter estimates, parameter standard errors, correlation coefficient of the fit, and autocorrelated structure parameter estimates are given in Table 5.2.

5.6 ARIMA(1,0,1)(0,0,1)7 Autocorrelation Function

The CTS and RTS algorithms require explicit formulas for the autocorrelation functions of the residual structure. In general, the autocorrelation function for a seasonal time series is complicated and depends on the order. Since the autocorrelation function for ARIMA(1,0,1)(0,0,1)7 is generally not given in the literature, this section will prove the computation in detail. The explicit formulas obtained in this section will be used directly to fit the model for the Iran trademark data discussed in Section 5.5.

To find the autocorrelation function of ARIMA(1,0,1)(0,0,1)7, note that the process includes a regular autoregressive (AR) component of order 1 and a moving average (MA) component of order 1, plus a seasonal MA component of order 1 and period 7. Therefore, the autocorrelation function has a seasonality lag at 6, 7, and 8, and a nonseasonality lag at 1 for a regular moving average and other lags of autocorrelation for a regular AR process. We split the process into nonseasonal and seasonal models using dummy variables. Let the time series

Z_t be the series following the ARIMA(1,0,1)(0,0,1)7 model. We write the non-seasonal part as the ARMA(1,1) process

$$(1 - \phi B)Z_t = (1 - \theta B)b_t \tag{5.4}$$

where B is a backward operator and b_t is the seasonality part of MA(1) and seasonality $s = 7$; that is,

$$b_t = (1 - \Theta B^7)a_t \tag{5.5}$$

with white noise process a_t, $Var(a_t) = \sigma_a^2$, and $Var(b_t) = \sigma_b^2$. Without loss of generality, assume that $E(z_t) = 0$; that is, the original series minus the mean, and $E(b_t) = 0$. The mixture of regular AR and seasonal MA leads to the ARIMA model:

$$(1 - \phi B)Z_t = (1 - \theta B)(1 - \Theta B^7)a_t \tag{5.6}$$

The regular AR(1) part of time series z_t, the MA(1)7 seasonal time series b_t, and their mixture can therefore be written as:

regular ARMA(1,1)	$z_t = \phi z_{t-1} + b_t - \theta b_{t-1}$ (5.7a)
seasonal MA(1)7	$b_t = a_t - \Theta a_{t-7}$ (5.7b)
mixture ARIMA(1,0,1)(0,0,1)7	$z_t = \phi z_{t-1} + a_t - \theta a_{t-1}$
	$\quad - \theta a_{t-7} + \theta \Theta a_{t-8}$ (5.7c)

The autocovariance function is defined as

$$\gamma_k = Cov(z_t, z_{t-k})$$
$$= E(z_t z_{t-k})$$

$$\zeta_k = Cov(b_t, b(t-k))$$
$$= E(b_t b_{t-k})$$

The autocorrelation function can be computed by $\rho_t = \gamma_t / \gamma_0$. Note that

$$E(z_t b_{t-j}) = E(z_{t-j} b_t) \begin{cases} \sigma_b^2 = \zeta_0 & j = 0 \\ 0 & j > 0 \end{cases}$$

$$E(b_t a_{t-j}) = E(z_{t-j} a_t) = \begin{cases} \sigma_a^2 = \gamma_0 & j = 0 \\ 0 & j > 0 \end{cases}$$

Equation (5.7a) leads to

$$E(b_{t-1} z_t) = \phi E(b_{t-1} z_{t-1}) + E(b_{t-1} b_t) - \theta E(b_{t-1}^2)$$
$$= (\phi - \theta)\zeta_0 \tag{5.8}$$

It is well known (see Wei (2006), Ch. 8) that for the MA(1) time series b_t with seasonality 7, only the autocorrelation function at lag 7 is nonzero. Therefore, for period j(7) for period $j = 1$ it is nonzero. This can be written as:

$$\zeta_k = \begin{cases} (\sigma_b^2) = (1 + \Theta^2)\sigma_a^2 & k = 0 \\ -\Theta\sigma_a^2 & k = 7 \\ 0 & elsewhere \end{cases} \tag{5.9}$$

which can be computed easily from (5.7b). To derive the autocovariance function for z_t, the regular AR(1) model 5.4 is mixed with irregular seasonal MA(1)7. From Equation 5.7a, z_t has lag differences 0,1 with b_t, so if we multiply it by z_{k-1} we have:

$$\gamma_k = \phi\gamma_{k-1} + E(z_{t-k}b_t) - \theta E(z_{t-k}b_{t-1}) \tag{5.10}$$

Specifically, when $k = 0$, using (5.8) we obtain:

$$\gamma_0 = \phi\gamma_1 + (1 - \theta\phi + \theta_1^2)\zeta_0$$

When $k = 1$ we have

$$\gamma_1 = \phi\gamma_0 - \theta\zeta_0$$

Solving the two linear equations at γ_0 and γ_2, we have

$$\gamma_0 = \frac{1 + \theta^2 - 2\theta\phi}{1 - \phi^2}\zeta_0 \tag{5.11a}$$

$$\gamma_1 = \frac{(\phi - \theta)(1 - \phi\theta)}{1 - \phi^2}\zeta_0 \tag{5.11b}$$

in which ζ_0 can be substituted from (5.9).

For higher lags, the autocorrelation at lags $k = 2, \ldots, 5$ can be obtained from (5.7a) and therefore (5.10). Then, since b_t is seasonal with seasonality 7 and period 1, which leads to Equation 5.7c, it can be seen there are lag differences at 1, 6, 7, and 8. For $k = 2, 3, 4, 5$ we have

$$\gamma_k = \phi^{k-1}\gamma_1$$

Unexpectedly, the lag $k = 4$ depends on the seasonality part because the seasonal part of the process on the right-hand side of (5.6) has 1, 6, 7, and 8 index differences. This will be used to compute the autocorrelation function for lag 6 onwards. Let $w_t = (1 - \phi B)z_t$ be the nonseasonal AR part. Equation 5.6 can then be written as:

$$w_t = (1 - \theta B)(1 - \Theta B^7)a_t \tag{5.12}$$

For computing autocovariances of lags 6, 7, and 8 we use the reverse formula of b_t, w_t and try to create a linear system of equations. We derive some primary formulas first. Using the autocovariance function (5.9) of b_t (5.7a), we have

$$E(z_{t-k}b_{t-j}) = \begin{cases} \phi^{j-k}\zeta_0 & j > k \\ \zeta_0 & j = k \\ 0 & j < k \end{cases}$$

for example $E(z_t b_{t-6}) = \phi^6 \zeta_0$. Writing the b_t formula (5.7a) in reverse we get

$$b_t = z_t - \phi z_{t-1} + \theta b_{t-1}$$

Since $\zeta_k = 0$ for $k \neq 7$, including $k = 6$, the above two equations result in

$$E(b_t b_{t-5}) = -\phi\gamma_4 + (1 + \phi^2)\gamma_5 - \phi\gamma_6 = 0$$
$$E(b_t b_{t-6}) = -\phi\gamma_5 + (1 + \phi^2)\gamma_6 - \phi\gamma_7 = 0$$
$$E(b_t b_{t-7}) = -\phi\gamma_6 + (1 + \phi^2)\gamma_7 - \phi\gamma_8 = -\Theta(1 - \theta^2)\sigma_a^2$$

therefore

$$\gamma_6 = \phi^5 \gamma_1$$
$$\gamma_7 = \phi^6 \gamma_1$$
$$\gamma_8 = \phi^7 \gamma_1 + \frac{\Theta(1-\theta^2)}{\phi}\sigma_a^2$$

For the other lags, the seasonal effect will be zero and both of the above approaches can be applied. Therefore, for $k = 9, \ldots, \infty$, using Equation 5.10 again, we conclude that

$$\gamma_9 = \phi\gamma_8$$
$$\gamma_k = \phi^{k-8}\gamma_8, k = 10, \ldots, \infty \tag{5.13}$$

In summary, we have proved the following theorem.

Theorem 5.1 The autocovariance function for time series $z_t \sim ARIMA$ $(1, 0, 1)(0, 0, 1)7$ is

$$\gamma_k = \begin{cases} \frac{1+\theta^2-2\theta\phi}{1-\phi^2}\zeta_0 & k = 0 \\ \frac{(\phi-\theta)(1-\phi\theta)}{1-\phi^2}\zeta_0 & k = 1 \\ \phi^{k-1}\gamma_1 & k = 2, 3, \ldots, 7 \\ \phi^7\gamma_1 + \frac{\Theta(1-\theta^2)}{\phi}\sigma_a^2 & k = 8 \\ \phi^{k-8}\gamma_8 & k = 9, 10, \ldots \end{cases} \tag{5.14}$$

with autocorrelation function

$$\rho_k = \gamma_k/\gamma_0, k = 1, \ldots \tag{5.15}$$

6

Outlier Detection in Nonlinear Regression

Previous chapters have discussed robust methods for fitting nonlinear regression models in null conditions, especially in the presence of outliers. Situations in which classical assumptions did not hold were also discussed. The methods presented reduce the impact of outlier data and lead to more modest estimates in the presence of null values. These estimators are applied to identify the outlier data in this chapter. Applying classical, nonrobust methods results in the false discovery of outliers. For example, Riazoshams et al. (2011) have shown some examples that mix least squares estimators with statistical measures, and which are not successful in identifying outliers.

Outlier detection is essential for practitioners because of the huge potential for misinterpretation in inference. High leverage points together with large errors (outliers) and residuals are responsible for masking and swamping of parameter estimates in linear regressions (see Habshah et al. (2009)). Moreover, masking and swamping breaches the assumptions of classical models. For example, Figure 3.1 shows how outliers not only have a masking and swamping effect on the model fit, but also misrepresent the model assumption properties, such as homogeneity and independence of errors. These facts mean that outlier detection methods need to be applied in cases where the classical assumptions are not also valid.

The outlier detection methods described in this chapter are based on computing several statistics combined with robust or classical parameter estimates. The theories and methods were developed by Riazoshams et al. (2011). The R-tools developed in the `nlr` package for following such an approach are explained in the second part of the book.

6.1 Introduction

The detection of outliers is essential because of they are responsible for a major interpretative problem in linear and nonlinear regression analysis. Many statistics practitioners have used residuals for the identification of

Robust Nonlinear Regression: with Applications using R, First Edition.
Hossein Riazoshams, Habshah Midi, and Gebrenegus Ghilagaber.
© 2019 John Wiley & Sons Ltd. Published 2019 by John Wiley & Sons Ltd.
Companion website: www.wiley.com/go/riazoshams/robustnonlinearregression

outliers. The use of residuals resulting from nonlinear least squares (OLS) estimates will give a misleading conclusion because the residuals are functions of leverages and true errors. According to Habshah et al. (2009), high leverage points together with large errors (outliers) and residuals are responsible the for masking and swamping of outliers in linear regressions. There are many good papers relating to the identaification of outliers in linear regression, for example Hadi (1992), Habshah et al. (2009), Cook and Weisberg (1982), Belsley et al. (1980), Anscombe and Tukey (1963), and the discussion of the properties of Atkinson's distance in Atkinson (1982, 1986).

However, little work has been carried out on the formulation of outlier identification measures in nonlinear regression. Cook and Weisberg (1982) and Fox et al. (1980) introduced a measure for the identification of outliers in a nonlinear model, based on the OLS method. Riazoshams et al. (2011) developed six outlier detection measures for nonlinear regression and showed that the studentized residuals and Cook distance, when combined with the robust estimator, can identify true outliers. Stromberg (1993) used graphical methods and robust estimators in nonlinear regression to detect outliers in the lakes data example. Motulsky and Brown (2006) proposed an algorithm for detecting outliers based on robust estimators. Vankeerberghen et al. (1995) used the least median squares (LMS) estimate combined with a genetic algorithm to detect outliers. Compared to linear regression, there been little research into outlier detection in nonlinear regression; in particular, to our knowledge, there has been no formal discussion of the identification of outliers in nonlinear regression when there are heteroscedastic variance errors.

However, it is evident that outliers have an adverse effect on the OLS estimates (see, for example, Habshah et al. (2009)). In this situation, we suspect that any measures based on OLS estimates are not efficient and this may cause swamping (false positive) and masking (false negative) effects.

In this chapter, the robust method of identification of outliers in a nonlinear model is discussed. The main idea is the linear approximation of a nonlinear model, and the method considers the gradient as the design matrix. Subsequently, detection techniques are formulated. Several outlier detection measures are developed that combine with OLS and MM-estimators. This work shows that, among the measures, only the studentized residual and Cook distance methods, combined with the MM-estimator, are consistently able to identify outliers correctly.

Some new definitions in outlier variances and detection in heteroscedastic variance cases are explored in this chapter. These techniques are employed in an `nlr` package to provide the R-tools for identifying atypical points.

6.2 Estimation Methods

In this chapter, the OLS and MM-estimators are utilized in the development of outlier detection measures. The intention is to use parameter estimates to compute statistical values that determine the influence of one observation on predictions of another value – an example would be the tangential leverage – and to incorporate them in statistical measures for outlier detection. These outlier detection measures are an alternative to the graphical technique. The identification of outliers by the graphical technique is more difficult or not possible when there is more than one independent variable in the model. Moreover, the use of classical outlier detection methods may give an incorrect result due to masking and swamping effects. In this situation, use of robust estimators and statistical measures is very important to correctly identify outliers in the data. In general it can be seen that the classical OLS estimate in conjunction with statistical measures is not capable of identifying bad data.

Point position again plays an important role. The point position can change the curvature and shape of a nonlinear model. In the heteroscedasticity of variance case, a new definition of "variance outliers" is explored in this chapter.

6.3   Point Influences

In this section, the influence of a data point is introduced. The definition will extend the concept of a point influence from linear to nonlinear regression by considering a linear approximation of the regression model. To avoid tangent plane approximations, the Jacobian leverage for a nonlinear regression is defined and its formula is extended to robust nonlinear regression. This is then combined with statistical measures to give measures for outlier detection.

To construct statistical measures for identifying influential observations, the intention is to measure the effect of one or more observations in predicting the response of another observation.

Consider the multiple linear regressions $Y = XB + \varepsilon$, where X is an $n \times p$ explanatory variable matrix, Y is an $n \times 1$ response vector, ε is an identically independent distributed (i.i.d.) error vector, n is the number of observations, and β is a p-dimensional unknown vector of coefficients. After the least squares estimates of the parameters $\hat{\beta}$ have been computed, the predicted value of the response variable can be written in the form of the hat matrix as follows:

$$\hat{Y} X \hat{\beta} = WY$$

where W is the hat matrix of

$$W = X(X^T X)^{-1} X^T$$

The elements of W are denoted by w_{ij}. It can be seen from this equation that the influence of the response values on the prediction depends on the values of w_{ij}. The equation can be rewritten as:

$$\hat{y}_i = w_{ii} y_i + \sum_{i \neq j; j=1}^{n} w_{ij} y_j \tag{6.1}$$

Hoaglin and Welsch (1978) suggested the direct use of w_{ij} as a diagnostic to identify high leverage points: if w_{ii} is large relative to the remaining terms, the fitted value $\hat{y}$ is dominated by the response $w_{ii}y_i$, so w_{ij} is interpreted as the amount of influence or leverage of $\hat{y}$ on y_i.

6.3.1 Tangential Plan Leverage

In nonlinear regression, a similar approach to (6.1) can achieved by linear approximation of the regression model around the true value of parameter, then replacing the design matrix in linear regression by the gradient of the function model. Let θ_0 be the true value of the parameter. Recall the gradient $\dot{\eta}$ of η from Equation 2.4, defined as

$$\dot{\eta}_{n\times p}(\theta) = \frac{\partial \eta(\theta)}{\partial \theta}$$

The linear approximation of the nonlinear regression model is then:

$$\eta(\theta) \simeq \eta(\theta_0) + \dot{\eta}(\theta_0)(\theta - \theta_0) \tag{6.2}$$

where $\dot{\eta}(\theta_0)$ is the gradient of η computed at the true value θ_0. The linear approximation (6.2) is a linear regression equation with design matrix $\dot{\eta}(\theta_0)$ and unknown parameter vector θ. It can be used to construct the least square estimates and predicted values, in a similar way to (6.1). The sum of the squared errors is defined as

$$S(\theta) = \|Y - \eta(\theta)\|^2$$
$$\approx \|Y - \eta(\theta_0) - \dot{\eta}(\theta_0)(\theta - \theta_0)\|^2$$

thus the OLS estimate is the answer to the minimization problem of the linear regression sum of squared errors with response values $[Y - \eta(\theta_0)] = \varepsilon$ and parameters $\beta = (\theta - \theta_0)$. Analogous to linear regression, the normal equation for the minimization problem is:

$$[\dot{\eta}^T(\theta_0)\dot{\eta}(\theta_0)]\beta = \dot{\eta}^T(\theta_0)\varepsilon \tag{6.3}$$

The OLS estimate of β is:

$$\hat{\beta} = \hat{\theta} - \theta_0$$
$$= [\dot{\eta}^T(\theta_0)\dot{\eta}(\theta_0)]^{-1}\dot{\eta}^T(\theta_0)\varepsilon$$

In practice $\dot{\eta}(\theta_0)$ is not known and has to be replaced by the estimate $\dot{\eta}(\hat{\theta})$, which is different from linear regression. The least squares estimate of constant variance σ is the answer to the equation

$$(n - p)\hat{\sigma} = S(\hat{\theta})$$

The predicted approximation (6.2) and the OLS estimate result in

$$\eta(\hat{\theta}) - \eta(\theta_0) = \dot{\eta}(\theta_0)[\dot{\eta}^T(\theta_0)\dot{\eta}(\theta_0)]^{-1}\dot{\eta}^T(\theta_0)\varepsilon$$

The hat matrix for nonlinear regression, referred to as the tangent plane leverage, is therefore defined as:

$$H = \dot{\eta}(\theta_0)[\dot{\eta}^T(\theta_0)\dot{\eta}(\theta_0)]^{-1}\dot{\eta}^T(\theta_0) \tag{6.4}$$

The leverage matrix in nonlinear regression plays a similar role to the hat matrix W in linear regression.

6.3.2 Jacobian Leverage

"The Jacobian leverage vector measures the magnitude of the derivative of each fitted value with respect to the mth response value."
(St. Laurent and Cook (1992)

The gradient usage involves the linear approximation of the nonlinear model in a single step of the quasi-Newton algorithm to find the least squares estimate (see Belsley et al. (1980)). However, the approximation of leverage in nonlinear regression will depend on the adequacy of the tangent plane approximation of the nonlinear model. For this reason, both a generalized leverage and a Jacobian leverage are defined by St. Laurent and Cook (1992, 1993). The latter is a direct definition of generalized leverage due to perturbation of one observation (Emerson et al. 1984). Emerson et al. (1984) and St. Laurent and Cook (1992) defined the Jacobian leverage (denoted by $\hat{J}_{ik}$) as the instantaneous rate of change for the ith fitted value with respect to the kth response.

The purpose of this section is to derive the generalized and Jacobian leverage of Wei et al. (1998) for the robust MM-estimator based on St. Laurent and Cook (1992) leverage. These derived leverages are then used in the development of the outlier detection measures.

Consider modifying the response Y by adding b to the mth element and writing the perturbed vector as $Y_{m,b} = Y + b\mathbf{a}_m$, where $\mathbf{a}_m$ is the mth standard basis vector in R^n. Denoting the least squares estimate of θ for the perturbed data by $\hat{\theta}_m(b)$, then the predicted response vector for the perturbed data is written as

$$\hat{Y}_{m,b} = \eta(\hat{\theta}_m(b)) \tag{6.5}$$

Emerson et al. (1984) defined the vector of the generalized leverage due to perturbation of the mth observation by b as

$$G(b; m) = \frac{1}{b}(\hat{Y}_{m,b} - \hat{Y}) \tag{6.6}$$

and the vector of the Jacobian leverage as

$$J(m) = \lim_{b \to 0} G(b; m) \tag{6.7}$$

where $\hat{Y} = \eta(\hat{\theta})$ is the predicted response for the unperturbed data.

St. Laurent and Cook (1992) derived the expression for the generalized and Jacobian leverage, (5.2) and (5.3) respectively, which is applicable to any sufficiently smooth nonlinear regression model and to a perturbation of the response vector in any direction in R^n. According to St. Laurent and Cook (1992), the generalized leverage can be written as

$$G(b;f) = \hat{V}\Delta_a^1 + \frac{1}{2}b(\hat{V}\Delta_a^2 - (\Delta_a^1)^T \hat{W}\Delta_a^1) \tag{6.8}$$

and the Jacobian leverage as

$$\hat{J} = \hat{V}(\hat{V}^T\hat{V} - [\hat{r}] \otimes [\hat{W}])^{-1}\hat{V}^T \tag{6.9}$$

where $\mathbf{a}$ is an arbitrary n-dimensional vector of unit length, for example $\mathbf{a}_m$, and $\hat{V}$ is an $n \times p$ gradient matrix computed using OLS, and $\hat{W}$ is an $n \times p \times p$ three-dimensional array of Hessian matrices. The ith face of W, denoted by W_i, is defined as:

$$W_i = W_i(\theta) = \frac{\partial \eta_i(\theta)}{\partial\theta\partial\theta^T}, i = 1, ..., n$$

and

$$\Delta_a^k = \frac{d^k}{db^k}\hat{\theta}_a(b)\Big|_{b=0} \tag{6.10}$$

for $k = 1, 2$; that is, Δ_a^k is the $p \times 1$ vector of the kth derivative of the elements of $\hat{\theta}_a(b)$ evaluated at $b = 0$ and $\hat{\theta}_a(b)$ is the OLS estimate of θ with perturbed data as a function of b and a, and is twice continuously differentiable in b (in the neighborhood of $b = 0$).

In Equation 6.9, the notation [ ] ⊗ [ ] represents column multiplication of a three-dimensional array, as defined by Bates and Watts (1980) (see Section 7.8). $\mathbf{r} = \mathbf{r}(\theta) = Y - \eta(\theta)$ is the n vector of residuals for a given value of θ. The elements of $\mathbf{r}$ are denoted as r_i.

The generalized and Jacobian leverage each have their own appeal as a measure of potential influence or leverage, so they can be considered as an alternative to the tangential plane leverage in the formulation of outlier detection measures for nonlinear regression. In the following sections, we attempt to derive the generalized and Jacobian leverages for the robust M- and MM-estimators.

6.3.3 Generalized and Jacobian Leverages for M-estimator

As discussed in Section 3.11, when σ is known, the M-estimator is defined as in Equation 3.14. Without loss of generality, when σ is unknown it is replaced by $\tilde{\sigma}$, which is the M-estimator computed in stage 3 of Algorithm 3.1. It is important to note that for the MM-estimator computed in Algorithm 3.1, the rho function $\rho_H(r/\sigma)$ is replaced by $\rho_H(r/(\tilde{\sigma}k_1))$.

Let $\tilde{\theta}_{\mathbf{a}}(b)$ be the vector valued function that gives the M-estimate of θ as a function of b and $\mathbf{a}_\bullet$. The tilde notation is used in general as any estimator computed at an M-estimate. In particular $\tilde{\theta}_{\mathbf{a}}(0)$ is the M-estimate of θ for unperturbed data. Assume $\tilde{\theta}_{\mathbf{a}}(b)$ is at least twice continuously differentiable in b (in the neighborhood of $b = 0$). The differentials are denoted with the same notation for the derivative as (6.10), but this time computed at an M-estimate.

To find the Jacobian leverage vector for the M-estimate, we follow the same approach as St. Laurent and Cook (1992), expanding $\eta(\tilde{\theta}_{\mathbf{a}}(b))$ about $b = 0$ via a second-order Taylor series. St. Laurent and Cook proved that the generalized leverage can be written as (6.8) for least squares. To derive the Jacobian leverage for the M-estimate, we first need to compute the derivatives $\Delta_a^k, k = 1, 2$, as defined in (6.10), for the M-estimate.

For perturbed data, the predicted response (6.5) of the M-estimate can be written as $\tilde{Y}_{a,b} = \eta(\tilde{\theta}_a(b))$, where the M-estimate $\tilde{\theta}_a(b)$ for the perturbed data is:

$$\tilde{\theta}_a(b) = \arg\min_\theta \sum_{i=1}^n \rho\left[\frac{y_i + ba_i - \eta_i(\theta_a(b))}{\sigma}\right] \tag{6.11}$$

where $\eta_i(\theta) = f(x_i; \theta)$. Without changing the notation of derivatives $\Delta_a^k, k = 1, 2$, to find the derivatives (6.10) of the M-estimate objective function, an equivalent normal equation to the OLS for the M-estimate can be computed from (6.11). Let us denote the kth normal equation by eq_k, defined as the derivative of the objective function (6.11) with respect to the kth element of vector $\theta = (\theta_1, ..., \theta_k, ..., \theta_p)$. The psi function ψ is defined as the derivative of the ρ function $\rho' = \psi$, and the second derivative of ψ or derivative of ψ', by $\zeta = \psi' = \rho''$. Then the kth equation, eq_k, can be written as

$$eq_k = \sum_{i=1}^n \left\{ \psi\left(\frac{r_{ia}(b)}{\sigma}\right) \frac{\partial \eta_i(\theta_a(b))}{\partial \theta_k} \right\} = 0, \; k = 1, ..., p \tag{6.12}$$

where $r_{ia}(b) = y_i + ba_i - \eta_i(\theta_\mathbf{a}(b))$. By differentiating (6.12) with respect to b and evaluating at $b = 0$, the derivatives $\Delta_a^k, k = 1, 2$ can be computed by

$$\frac{d}{db} eq_k = \sum_{i=1}^n \left\{ \psi'\left(\frac{r_{ia}(b)}{\sigma}\right) \left(\frac{a_i - V_{i\bullet}^T \Delta_a^1}{\sigma}\right) \frac{\partial \eta_i(\theta_a(b))}{\partial \theta_k} \right.$$
$$\left. + \psi\left(\frac{r_{ia}(b)}{\sigma}\right) \Delta_a^{1^T} w_{ik\bullet} \right\}, \quad k = 1, ..., p \tag{6.13}$$

The first term of the derivative with respect to b is:

$$\frac{d\eta_i(\theta_a(b))}{db} = \sum_{j=1}^{p} \frac{\partial \eta_i}{\partial \theta_j} \frac{d\theta_j}{db}$$

$$= [V_{i1}, ..., V_{ip}]_{1\times p} \Delta_a^1 \tag{6.14}$$

$$= V_{i\bullet} \Delta_a^1$$

and the second term is

$$\frac{d}{db} \frac{\partial \eta_i(\theta_a(b))}{\partial \theta_k} = \sum_{j=1}^{p} \frac{\partial^2 \eta_i}{\partial \theta_k \partial \theta_j^T} \frac{d\theta_j}{db}$$

$$= \Delta_a^{1^T} \begin{bmatrix} w_{ik1} \\ \vdots \\ w_{ikp} \end{bmatrix} \tag{6.15}$$

$$= \Delta_a^{1^T} w_{ik\bullet}$$

where the dotted subscript shows the vector split from the matrix or three-dimensional array in the dotted dimension; that is,

$$V_{i\bullet}^T = [V_{i1}, ..., V_{ip}]_{1\times p}, i = 1, ..., n,$$

is the gradient for ith data point, and $w_{ikg}^T = [w_{ik1}, ..., w_{ikp}]$ is the Hessian matrix at the ith data point and column k. Solving equation (6.13) gives the value of Δ_a^1. Note that $V_{i\bullet}^T \Delta_a^1 = \Delta_a^{1(T)} V_{i\bullet}$. After some algebraic operations we get

$$\frac{d}{db} eq_k - \sum_{i=1}^{n} \left\{ \zeta_i(b) \frac{a_i}{\sigma} V_{ik} - \frac{1}{\sigma} \zeta_i(b) \Delta_a^{1(T)} V_{i\bullet} V_{ik} + \psi_i(b) \Delta_a^{1(T)} w_{ik\bullet} \right\}$$

$$= \frac{1}{\sigma} \sum_{i=1}^{n} a_i \zeta_i(b) V_{ik} - \sum_{i=1}^{n} \Delta_a^{1^T} \left\{ \frac{1}{\sigma} \zeta_i(b) V_{i\bullet} V_{ik} - \psi_i(b) w_{ik\bullet} \right\} = 0$$

$$\Rightarrow \frac{1}{\sigma} a^T [\zeta \circ V_{\bullet k}] = \Delta_a^{1(T)} \left[\frac{1}{\sigma} (\zeta \circ V_{\bullet k})^T V - \psi^T W_{\bullet k\bullet} \right] \forall k$$

Writing this in matrix form, the value of Δ_a^1 is the solution of the following equation:

$$\frac{1}{\sigma} [\zeta \circ V_{\bullet j}]^T a = \left(\frac{1}{\sigma} [\zeta \circ \tilde{V}_{\bullet j}]^T V - [\psi^T] \otimes [W] \right) \Delta_a^1 \tag{6.16}$$

The product $\circ$ is the Hadamard elementwise product and $\zeta \circ V_{\bullet j}$ is the Hadamard product in columns of V; that is,

$$[\zeta \circ V_{\bullet j}] = \begin{bmatrix} \zeta_1 V_{11} & \cdots & \zeta_1 V_{1p} \\ \vdots & \vdots & \vdots \\ \zeta_n V_{n1} & \cdots & \zeta_n V_{np} \end{bmatrix} = \begin{bmatrix} \zeta_1 \circ V_{\bullet 1} & \cdots & \zeta_1 \circ V_{\bullet p} \end{bmatrix}$$

Calculating the Jacobian leverage by finding the limit when $b \to 0$ gives the value of $T = \tilde{V}\Delta_a^1$ (the tilde denotes it is computed at M-estimate $\tilde{\theta}$), and substituting the solution of Δ_a^1 from (6.16) yields the robust Jacobian leverage for the M-estimate as follows:

$$\tilde{T} = \frac{1}{\sigma}\tilde{V}\left(\frac{1}{\sigma}[\zeta \circ \tilde{V}_{\bullet j}]^T \tilde{V} - [\psi^T] \otimes [\tilde{w}]\right)^{-1}[\zeta \circ \tilde{V}_{\bullet j}]^T \tag{6.17}$$

where $[\zeta \circ \tilde{V}_{\bullet j}]$ is the Hadamard product of ζ in columns of $\tilde{V}$; that is,

$$[\zeta \circ \tilde{V}_{gj}] = [\zeta \circ \tilde{V}_{g1}, ..., \zeta \circ \tilde{V}_{gp}]$$

For OLS the robust Jacobian leverage and the Jacobian leverage give the same result. In OLS the ρ function is $\rho(t) = t^2$ and $\psi(t) = 2t$, $\zeta(t) = 2$, where $t = r_i/\sigma$. Substituting these values gives

$$\tilde{T} = \frac{1}{\sigma}\tilde{V}\left(\frac{1}{\sigma}[2 \circ \tilde{V}_j]^T \tilde{V} - [\frac{1}{\sigma}2r^T] \otimes [\tilde{w}]\right)^{-1}[2 \circ \tilde{V}_{\bullet j}]^T$$
$$= \frac{1}{\sigma}\hat{V}\left(\frac{\sigma}{2}\right)(\hat{V}^T\hat{V} - [r^T] \otimes [\hat{w}])^{-1}\, 2\hat{V}^T$$
$$= \hat{V}(\hat{V}^T\hat{V} - [r^T] \otimes [\hat{w}])^{-1}\hat{V}^T$$

which is equal to the Jacobian leverage (6.9).

In the next section we will incorporate the new robust Jacobian leverage in six outlier detection measures. The residuals for the outlier observations of the MM-estimates are very large. Consequently, the Jacobian leverage in the equation and the $[\mathbf{r}]$ vector become much bigger for outlier observations than for other values. As a result, computing the inverse $(\hat{V}^T\hat{V} - [\hat{\mathbf{r}}] \otimes [\hat{W}])^{-1}$ raises a singularity problem due to the rounding error.

However, the robust Jacobian leverage does not have this problem since the residual vector $[\mathbf{r}]$ is replaced by a function ψ that will downplay the effect of outliers. The singularity problem is common in all computational processes, since residuals for outliers are large.

6.4 Outlier Detection Measures

Linear regression uses the hat matrix W as an influence-detection tool and creates several statistical measures for outlier detection. The development of such measures is based on classical estimates. In this chapter, the tangent plane leverage matrix H of Equation 6.4 is used in the formulation of the outlier detection measure statistics in nonlinear regression. The proposed measures are also based on robust MM-estimates.

A common way of developing an influence detection method is to re-fit a model by deleting a special case or a set of cases. The amount of change of various statistics – the parameter estimates, predicted likelihoods, residuals, and so on – is observed for a recalculated measure with the ith data point removed. The notation $-i$ is used for each removed observation. It is important to point out that the OLS, M- and MM-estimators are used to estimate the parameters of the nonlinear regression. Subsequently, the respective estimates are used in the computation of the influence measures. Several outlier measures are briefly discussed as follows.

In general, we do not recommend deleting data points in a nonlinear regression, except when an explicit formula exists, because the position of the points affects the shape of the model curve and has an unpredictable effect on rounding errors, convergence iteration, and model fit. Although some functions using point deletion are provided in the `nlr` package, the approach is not recommended and the functions are used in this book for comparison purposes, and not for inference.

6.4.1 Studentized and Deletion Studentized Residuals

This measure (hereafter referred to as t_i) is used for identifying outliers. Suppose h_{ii} is the diagonal of leverage matrix H based on the gradient in Equation 6.4. The studentized residual and the deleted studentized (standardized) residuals are defined as (see Srikantan (1961) and Anscombe and Tukey (1963)):

$$t_i = \frac{r_i}{\hat{\sigma}\sqrt{1 - h_{ii}}} \tag{6.18}$$

and

$$d_i = \frac{r_i}{\hat{\sigma}\sqrt{1 - h_{ii}}}$$

respectively, where $\hat{\sigma}_{(-i)}$ is the estimated standard deviation in the absence of the ith observation. The residuals, denoted by $r_i = y_i - f(x_i; \hat{\theta})$, are obtained from the OLS and MM-estimates. The ith observation is considered as an outlier if $|t_i|$ or $|d_i| > 3$.

The studentized (standardized) residual (6.18) is defined as the residual divided by its standard error. Note that:

$$\mathbf{Y} - \eta(\theta_0) = \varepsilon$$

and the vector of residuals can be written by linear approximation as:

$$\mathbf{r} = \mathbf{Y} - \eta(\hat{\theta}) = \mathbf{Y} - \eta(\theta_0) - \dot{\eta}(\theta_0)(\hat{\theta} - \theta_0)$$
$$= (I - H)\varepsilon$$

Assuming the variance of error is homogeneous ($Var(\varepsilon) = \sigma^2 I$), then the variance of residuals is equal to

$$Var(\mathbf{r}) = \sigma^2(I - H).$$

Note that the standard error σ is unknown and will be replaced by the least squares estimate $\hat{\sigma}$. Let $r_i = y_i - \eta_i(\hat{\theta})$ be the residual of the ith observation, $Se(r_i)$ the standard error of the residual, and h_{ii} the ith diagonal of the hat matrix H. The standardized residual (denoted by t_i) can then be computed by

$$t_i = r_i / Se(r_i) = \frac{r_i}{\hat{\sigma}\sqrt{(1 - h_{ii})}} \tag{6.19}$$

Studentized (standardized) residuals are used to detect outliers by many practitioners, but masking and swamping effects should be taken into account. The OLS estimate is not robust and will be affected by a single outlier, so the residuals of outliers will be masked and residuals of good data will be swamped.

6.4.2 Hadi's Potential

Hadi (1992) proposed the potential p_{ii} to detect high leverage points or large residuals:

$$p_{ii} = \frac{h_{ii}}{1 - h_{ii}}$$

The cut-off point for p_{ii} was proposed to be $Median(p_{ii}) + cMAD(p_{ii})$, where MAD is the mean absolute deviation, defined by:

$$MAD(p_{ii}) = Median[p_{ii} - Median(p_{ii})]/0.6745$$

and c is an appropriately chosen constant such as 2 or 3.

6.4.3 Elliptic Norm (Cook Distance)

The Cook distance (CD), defined by Cook and Weisberg (1982), is used to assess influential observations. An observation is influential if the value of CD is greater than 1. Cook and Weisberg defined CD as

$$CD_i(V^T V, p\hat{\sigma}^2) = \frac{(\theta - \hat{\theta}_{(-i)})^T (V^T V)(\theta - \hat{\theta}_{(-i)})}{p\hat{\sigma}^2}$$

where $\hat{\theta}_{(-i)}$ are the parameter estimates when the ith observation is removed. When $\hat{\theta}_{(-i)}$ is replaced by the linear approximation (see Cook and Weisberg (1982) and Fox et al. (1980)), this norm changes to

$$CD_i(V, p\sigma^2) = \frac{t_i^2}{p} \frac{h_{ii}}{1 - h_{ii}} \tag{6.20}$$

where t_i and p are the studentized residual and the number of parameters in the model respectively, with cut-off point equal to 1, which is the expectation of the largest segment of the 50% confidence ellipsoid of parameter estimates.

6.4.4 Difference in Fits

Difference in fits (*DFFITS*) is another diagnostic measure used in measuring influence. It was defined by Belsley et al. (1980). For the ith observation, *DFFITS* is defined as

$$DFFITS_i = \left(\sqrt{\frac{h_{ii}}{1 - h_{ii}}} \right) |d_i|$$

where d_i is the deleted studentized residual. The authors considered an observation to be an outlier when *DFFITS* exceeds a cut-off point of $2\sqrt{p/n}$.

6.4.5 Atkinson's Distance

The Atkinson distance for observation i (C_i) was developed by Anscombe and Tukey (1963) and is used to detect an influential observation (Atkinson 1982, 1986) for the discussion of Atkinson's property. Atkinson (1982) defined the Atkinson distance as

$$C_i = \left(\sqrt{\frac{n - p}{p} \frac{h_{ii}}{1 - h_{ii}}} \right) |d_i|$$

where d_i are the deleted studentized residuals. He suggested a cut-off value of 2.

6.4.6 DFBETAS

DFBETAS is a measure of how much an observation has effected the estimate of a regression coefficient (there is one DFBETA for each regression coefficient, including the intercept). For nonlinear regression, replacing the design matrix $\mathbf{X}$ with gradient $\dot{\eta}$, DFBETAS for the ith data and jth parameters can be computed as:

$$DFBETAS_{j(i)} = \frac{\hat{\beta}_j - \hat{\beta}_{j(i)}}{\sqrt{\hat{\sigma}^2_{(i)} [\dot{\eta}^T(\theta_0)\dot{\eta}(\theta_0)]^{-1}_{jj}}} \tag{6.21}$$

For small/medium datasets, an absolute value of 1 or greater is suspicious. For a large dataset: absolute value larger than $2/\sqrt{n}$ are considered highly influential.

In nonlinear regression, the parameter effects are different from linear regression, and deleting a single point affects the curve of the model. We therefore do not recommend using deletion methods. In particular, the effect on a parameter of deleting a point is not known, so we will not discuss this measure further.

6.4.7 Measures Based on Jacobian and MM-estimators

As discussed in previous sections, the tangential leverage, OLS, and MM-estimators are incorporated in the six outlier detection measures for identifying outliers. Following on from this, we incorporated the nonrobust Jacobian and robust Jacobian leverage with OLS and MM-estimators in the same way. The main motivation was to first compute the robust MM-estimate $\hat{\theta}$, followed by the residuals and the Jacobian leverage $\hat{J}$. Then the robust Jacobian leverage, as derived in Equation 6.17, was computed. Both the Jacobian leverage and the robust Jacobian leverage were then incorporated in the six outlier detection measures. This is a direct generalization of linear regression using the hat matrix, but as can be seen, since the robust estimates downgrade the effect of outliers, their influence on the parameter estimates is not large.

In fact, the robustified Jacobian leverage really measures the influence of points in the prediction. Since influence of outliers is reduced by the robust estimator, and because it uses robust estimated values, robust Jacobian leverage cannot distinguish the outliers from other data points. In this situation, it may not be a good alternative to tangential plane leverage for formulating an algorithm for outlier detection incorporating the six detection measures.

Consider the six outlier measures that were discussed in Sections 6.4.1–6.4.5. Let h_{ii} be the diagonal elements of the tangential plane leverage (Equation 6.4). Let $\hat{J}_{ii}$ be the diagonal elements of Jacobian leverage for the least squares estimate in Equation 6.9, and $\tilde{T}_{ii}$ be the diagonal elements of the robust Jacobian leverage (Equation 6.17). In the presence of outliers, the residual computed from the MM-estimates, denoted as vector $[\mathbf{r}]$ in the Jacobian leverage in Equation 6.9, is very large. Subsequently, the inverse of matrix $(\hat{V}^T \hat{V} - [\hat{\mathbf{r}}] \otimes [\hat{W}])^{-1}$ sometimes cannot be calculated due to singularity. As an alternative, the robust Jacobian leverage is utilized.

Super leverage refers to a point that has leverage value greater than 1. In linear regression, the leverage values are less than 1, but in nonlinear regression, leverage values greater than 1 can occur. For super leverages, if $h_{ii} > 1$ then $0 > 1 - h_{ii}$ and the square of this value cannot be computed. In this situation, the robust Jacobian leverage may be applied.

6.4.8 Robust Jacobian Leverage and Local Influences

Several authors have discussed the fact that the leverage, local influence, and model influence are related to each other (see for example Cook (1986), St. Laurent and Cook (1992, 1993), Cook et al. (1986), Belsley et al. (1980), and Bates and Watts (2007)).

As explained in the previous chapter, in linear regression the hat matrix is used to compute the influence of one observation in estimating the response at other points. In nonlinear regression, the same inference is achieved by replacing the design matrix with a gradient of model function, called the tangential plane leverage matrix. In this section, an alternative way of formulating a robust outlier detection measure is introduced.

We show here how to use the proposed robust Jacobian leverage to identify an influential observation. The intention is to calculate the difference between the robust Jacobian leverage (Equation 6.17) and the tangent plane leverage (Equation 6.4). The nonrobust Jacobian leverage is not discussed because, as mentioned before, it has many computational problems in the presence of outliers.

As explained before, if the diagonal values of tangent plane leverage H (say h_{ii}) are large relative to the remaining terms, the fitted value of the ith ($\hat{y}_i$) point is more influenced by the ith response y_i. Nevertheless, this situation is valid only when OLS adequately estimates the model in a linear approximation (see Bates and Watts (1980, 2007) and Belsley et al. (1980) for a discussion of curvature). The generalized leverage (Equation 6.6) and Jacobian leverage (Equation 6.7) each have their appeal as a measure of potential influence or leverage. The Jacobian leverage vector measures the magnitude of the derivative of each fitted value with respect to the mth response value.

Tangent plane leverage values depend on the adequacy of the inear approximation (Equation 6.2). On the other hand, the Jacobian leverage, and consequently the robust Jacobian leverage, directly measure the effect of the perturbation of one observation in a prediction, so the difference between these measures can be a way of identifying influence values (see, for example, St. Laurent and Cook (1992)). However, as explained before, in some cases the nonrobust Jacobian leverage cannot be computed in the presence of outliers. In addition, by definition, the derived robust Jacobian leverage is an exact mathematical formula, so it is more reliable. Since the robust estimate downgrades the effect of outliers, its values are smaller than the tangential leverage in the presence of outliers. Therefore, the difference between the robust Jacobian leverage and the tangential leverage maybe large (negative) for extreme points. This difference is denoted by $DLEV$ and is defined as

$$DLev_i = t_{ii} - h_{ii}; i = 1, ..., n \tag{6.22}$$

where t_{ii} are the diagonal elements of the robust Jacobian leverage $\tilde{T}$ (Equation 6.17) and h_{ii} are the diagonal elements of the tangent plane leverage matrix H (Equation 6.4). This quantity increases negatively with an increase in magnitude of the extreme points. There is no finite upper bound for $DLev_i$ and the theoretical distribution of $DLVE$ is not tractable. Nonetheless this does not cause any difficulties in obtaining suitable confidence-bound type cut-off points for them. Following the proposal of Hadi (1992), which was subsequently used by many others (see Imon (2002, 2005) and Habshah et al. (2009)), a form analogous to a confidence bound for a location parameter was proposed. We consider $DLev_i$ to be large if

$$DLev_i < Median(DLev_i) - c \times MAD(DLev_i)$$

where

$$MAD(DLev_i) = Median\{|DLev_i - Median(DLev_i)|\}/0.6745$$

and c is an appropriate chosen constant such as 2 or 3.

6.4.9 Overview

Note that the applied diagonal of the hat matrix h_{ii} depends on an unknown parameter and hence the estimate of parameter θ should be substituted. The possible choices are the classical OLS or robust MM-estimates. Riazoshams et al. (2011) extended the above six measures for nonlinear regression and showed that studentized residuals and the Cook distance, when combined with the robust MM-estimator, can identify the outlying points. In the next section we display results for only these two measures in most cases. The results in this chapter are slightly different from those of Riazoshams et al. (2011) because the programs used have improved in recent years.

The argument of Riazoshams et al. (2011) was based on simulation. It is not clear yet that the two measures fail in the examples, but intuitively we expect they will give an appropriate result in most cases. In addition, they will not work properly in the presence of heteroscedasticity of variance or autocorrelated regression errors. Therefore we have further developed them so that they can handle heteroscedastic error variances.

R-tools are provided in the nlr package that will compute and display the outlier detection measures in graphical format. To avoid recomputation for each measure, the implemented program computes all the measures, including measures that depend on deleting a data point, but users should remember that only the studentized residual and Cook distance measures work properly. Other measures can be used for other purposes and require more study. Furthermore, deleting data points has an unpredictable effect since the point position is important in nonlinear regression.

In general, we not recommend deleting data points in nonlinear regression because of the point position effect because of the computational cost, although some functions provided in nlr use point deletion. In the following sections the measures that are not based on deleting points are discussed and, since the studentized residual and Cook distance measures are the measures that successfully identify outliers when combined with the robust estimate, these and the MM-estimate will be considered for most of the inferences.

6.5   Simulation Study

A simulation study was performed to assess the capability of measures for identifying outlying points. The simulated data are shown in Table A.9.

The simulated value from the logistic model is based on the following function:

$$y_i = \frac{a}{1 + b.e^{-cx_i}} + \varepsilon_i$$

where $\theta = (a, b, c)$ are the parameters of the model. This model is chosen to mimic the real-life chicken-growth data set in Chapter 4.9 (see Riazoshams and Miri (2005) and Riazoshams and Midi (2009)). In this simulation study, three parameters are considered: $a = 2575, b = 41$, and $c = 0.11$. The residuals are generated from a normal distribution with mean zero and standard deviation $\sigma = 70$. The x_i are generated from a uniform distribution on $[3, 50]$ with a sample size of 20. Three different cases of contamination are considered.

Case A: The first good datum point (x_1, y_1) is replaced with a new value y_1 equal to $y_1 + 1000$. The graph is shown in Figure 6.1a for the OLS and MM-estimates.

Case B: The sixth, seventh and eighth data points are replaced with their corresponding y values increased by 1000 units. The graph is shown in Figure 6.1b for the OLS and MM-estimates.

Case C: Six high leverage points were created by replacing the last six observations with (x,y) pair values: (90, 6500), (92, 6510), (93, 6400), (93, 6520), (90, 6600), (94, 6600). The graph is shown Figure 6.1c for OLS and MM-estimates.

Let us first focus on Case A, where the first observation is an outlier. It can be seen from Tables 6.1 and 6.2 that all methods fail to detect the single outlier when we consider the nonrobust and robust Jacobian leverages, except the t_i-OLS method based on H and $\hat{J}$, the t_i-MM method based on $H, \hat{J}$, and $\tilde{T}$, and the CD_i-MM method based on H and $\hat{J}$.

The results also show that the p_{ii}-MM method based on $\hat{J}$ and the p_{ii}-MM method based on $\tilde{T}$ mask the outlier point and swamp the other points. Here, the robust Jacobian leverage $\tilde{T}$ shows less influence of outliers in estimating the parameters of a model.

For Case B, Tables 6.3 and 6.4 show the results for three outliers, when using the least squares estimate based on nonrobust Jacobian leverage, and the MM-estimate based on robust Jacobian leverage $\tilde{T}$. For this case, the nonrobust Jacobian leverage $\hat{J}$ cannot be computed, as explained before. The result shows that only t_i-MM based on H, $\tilde{T}$, and CD_i based on H are able to identify the three outliers correctly. Other outlier measures fail to identify

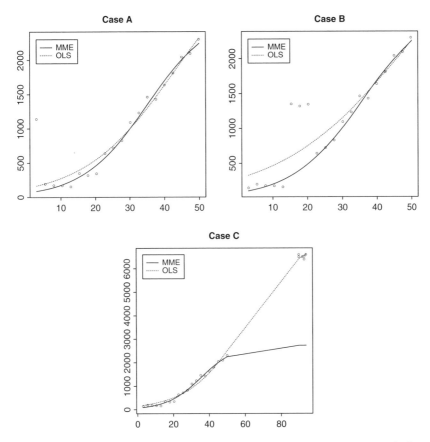

Figure 6.1 Simulated outliers. Case A: one outlier; Case B: three outliers; Case C: six high leverage points.

even a single outlier. For instance, p_{ii}-OLS and the tangent leverage masked the three outliers and swamped the other observations (data case numbers 19 and 20). Swamping can be observed for other measures in different mixing situations.

It is important to note that in Table 6.4 several outlier detection measures cannot be computed because of super leverage points; that is, the leverage values are greater than 1 (St. Laurent and Cook (1992)). However, this problem could be solved by using the robust Jacobian leverage, as shown in the table. Using the robust Jacobian leverage and t_i-MM, the first observation was successfully identified as an outlier.

Table 6.4 is based on the MM, classical and robust Jacobian leverage $\tilde{T}, \hat{J}$. It is worth mentioning here that the nonrobust Jacobian leverage $\hat{T}$ defined in

Table 6.1 Measures based on OLS estimates for a data set with one outlier (Case1): mixture of OLS with tangential leverage and Jacobian leverage.

Measure	OLS estimate, tangential leverage						OLS estimate, Jacobian leverage					
	t_i	d_i	CD_i	P_{ii}	$DFFITS_i$	C_i	t_i	d_i	CD_i	P_{ii}	$DFFITS_i$	C_i
cutf	3.000	3.000	1.000	0.287	0.775	2.000	3.000	3.000	1.000	0.308	0.775	2.000
1	3.908	0.781	0.597	0.070	0.207	0.492	3.978	0.795	0.758	0.109	0.258	0.614
2	0.370	0.046	0.063	0.086	0.014	0.032	0.378	0.047	0.080	0.135	0.017	0.040
3	−0.735	−0.124	0.136	0.102	0.040	0.095	−0.754	−0.128	0.175	0.161	0.050	0.119
4	−0.525	−0.097	0.104	0.117	0.033	0.079	−0.541	−0.100	0.134	0.185	0.042	0.099
5	0.114	0.022	0.024	0.128	0.008	0.019	0.118	0.023	0.031	0.201	0.010	0.024
6	0.007	0.001	0.001	0.134	0.001	0.001	0.007	0.001	0.002	0.206	0.001	0.002
7	−0.471	−0.087	0.099	0.132	0.031	0.075	−0.484	−0.089	0.123	0.195	0.038	0.091
8	−0.661	−0.106	0.136	0.126	0.038	0.090	−0.675	−0.109	0.162	0.173	0.044	0.105
9	−0.643	−0.089	0.128	0.119	0.031	0.073	−0.651	−0.090	0.144	0.146	0.034	0.081
10	−0.184	−0.022	0.036	0.118	0.008	0.018	−0.184	−0.022	0.038	0.126	0.008	0.019
11	−0.229	−0.027	0.047	0.128	0.010	0.023	−0.229	−0.027	0.046	0.122	0.009	0.023
12	0.021	0.003	0.005	0.150	0.001	0.003	0.020	0.003	0.004	0.139	0.001	0.002
13	0.402	0.067	0.098	0.178	0.028	0.067	0.401	0.067	0.097	0.174	0.028	0.067
14	0.040	0.008	0.010	0.200	0.004	0.009	0.040	0.008	0.011	0.211	0.004	0.009
15	0.480	0.106	0.124	0.202	0.048	0.113	0.485	0.107	0.134	0.228	0.051	0.121
16	0.571	0.120	0.141	0.184	0.052	0.123	0.579	0.122	0.155	0.215	0.056	0.133
17	−0.082	−0.015	0.020	0.171	0.006	0.015	−0.083	−0.015	0.021	0.192	0.007	0.016
18	−0.033	−0.007	0.009	0.211	0.003	0.008	−0.033	−0.007	0.009	0.213	0.003	0.008
19	−0.538	−0.195	0.194	0.389	0.121	0.289	−0.535	−0.193	0.187	0.369	0.118	0.282
20	0.021	0.020	0.012	1.000	0.020	0.047	0.021	0.020	0.012	0.969	0.019	0.046

cutf: abbreviation for cut of point.

Table 6.2 Measures based on MM-estimates for a data set with one outlier (Case 1): mixture of MM with tangential leverage, Jacobian leverage, and robust Jacobian leverage.

Measure	MM-estimate, tangential leverage							MM-estimate, Jacobian leverage						MM-estimate, robust Jacobian leverage					
	t_i	d_i	CD_i	p_{ii}	$DFFITS_i$	C_i	$DLEV$	t_i	d_i	CD_i	p_{ii}	$DFFITS_i$	C_i	t_i	d_i	CD_i	p_{ii}	$DFFITS_i$	C_i
1	11.451	0.000	1.178	0.032	0.000	0.000	−0.031	9.078	0.000		−0.352			11.274	0.000	0.000	0.000	0.000	0.000
2	1.829	0.000	0.226	0.046	0.000	0.000	−0.044	1.342	0.000		−0.437			1.788	0.000	0.000	0.000	0.000	0.000
3	−1.103	0.000	0.162	0.065	0.000	0.000	0.011	−0.743	0.000		−0.517			−1.109	0.000	0.178	0.077	0.000	0.000
4	−0.432	−0.049	0.074	0.087	0.014	0.034	0.014	−0.267	−0.030		−0.586			−0.435	−0.049	0.081	0.105	0.016	0.037
5	1.416	0.000	0.274	0.113	0.000	0.000	0.018	0.808	0.000		−0.638			1.430	0.000	0.304	0.135	0.000	0.000
6	1.161	0.000	0.248	0.137	0.000	0.000	0.020	0.625	0.000		−0.670			1.174	0.000	0.274	0.163	0.000	0.000
7	−0.204	−0.043	0.046	0.154	0.017	0.041	0.021	−0.108	−0.023		−0.679			−0.206	−0.044	0.051	0.182	0.019	0.044
8	−0.841	0.000	0.194	0.159	0.000	0.000	0.018	−0.458	0.000		−0.657			−0.850	0.000	0.211	0.184	0.000	0.000
9	−0.967	0.000	0.219	0.154	0.000	0.000	0.013	−0.581	0.000		−0.584			−0.974	0.000	0.233	0.172	0.000	0.000
10	0.085	0.008	0.019	0.149	0.003	0.007	0.006	0.061	0.006		−0.414			0.085	0.008	0.019	0.156	0.003	0.007
11	−0.358	−0.055	0.081	0.154	0.021	0.051	0.000	−0.318	−0.049		−0.088			−0.358	−0.055	0.081	0.154	0.021	0.051
12	0.008	0.002	0.002	0.178	0.001	0.002	−0.002	0.008	0.002	0.002	0.175	0.001	0.002	0.008	0.002	0.002	0.176	0.001	0.002
13	0.765	0.163	0.203	0.212	0.075	0.179	0.001	0.680	0.145		−0.042			0.766	0.164	0.204	0.214	0.076	0.180
14	−0.512	−0.121	0.142	0.230	0.058	0.139	0.006	−0.383	−0.091		−0.311			−0.514	−0.122	0.145	0.239	0.059	0.141
15	0.616	0.140	0.165	0.216	0.065	0.155	0.009	0.428	0.097		−0.414			0.619	0.141	0.171	0.229	0.067	0.160
16	0.919	0.001	0.227	0.184	0.000	0.001	0.007	0.670	0.000		−0.371			0.923	0.001	0.235	0.194	0.000	0.001
17	−0.702	−0.130	0.168	0.173	0.054	0.129	0.003	−0.592	−0.110		−0.165			−0.703	−0.130	0.171	0.177	0.055	0.130
18	−0.224	−0.047	0.061	0.225	0.022	0.053	−0.001	−0.221	−0.047	0.055	0.186	0.020	0.049	−0.224	−0.047	0.061	0.224	0.022	0.053
19	−1.155	−0.001	0.417	0.391	0.000	0.001	0.000	−1.061	−0.001	0.254	0.172	0.000	0.001	−1.156	−0.001	0.418	0.392	0.000	0.001
20	1.092	1.120	0.557	0.781	0.990	2.356	0.008	0.691	0.708		−0.287			1.100	1.128	0.570	0.806	1.006	2.394
cutf	3.000	3.000	1.000	0.377	0.775	2.000	−0.023	3.000	3.000	1.000	0.544	0.775	0.000	3.000	3.000	1.000	0.351	0.775	2.000

Empty cells are shown where values cannot be calculated because of super leverage or the singularity matrix. cutf: abbreviation for cut of point.

Table 6.3 Measures based on OLS estimates for a data set with three outliers (Cases 6, 7, and 8): mixture of OLS with tangential leverage and Jacobian leverage.

Measure	OLS estimate, tangential leverage						OLS estimate, Jacobian leverage					
	t_i	d_i	CD_i	p_{ii}	$DFFITS_i$	c_i	t_i	d_i	CD_i	p_{ii}	$DFFITS_i$	c_i
cutf	3.000	3.000	1.000	0.229	0.775	2.000	3.000	3.000	1.000	0.226	0.775	2.000
1	−0.662	−0.093	0.155	0.165	0.038	0.090	−0.653	−0.092	0.138	0.134	0.034	0.081
2	−0.299	−0.040	0.070	0.165	0.016	0.039	−0.295	−0.039	0.062	0.134	0.015	0.035
3	−1.170	−0.162	0.270	0.160	0.065	0.154	−1.155	−0.160	0.241	0.131	0.058	0.139
4	−1.046	−0.132	0.233	0.149	0.051	0.121	−1.035	−0.130	0.211	0.124	0.046	0.111
5	−0.604	−0.068	0.128	0.135	0.025	0.059	−0.599	−0.067	0.117	0.115	0.023	0.055
6	2.338	0.301	0.466	0.119	0.104	0.248	2.324	0.299	0.436	0.106	0.098	0.233
7	1.972	0.222	0.371	0.106	0.072	0.172	1.965	0.221	0.357	0.099	0.070	0.166
8	1.836	0.194	0.333	0.099	0.061	0.145	1.834	0.194	0.330	0.097	0.060	0.144
9	−1.136	−0.120	0.206	0.099	0.038	0.090	−1.137	−0.120	0.209	0.102	0.038	0.091
10	−0.760	−0.088	0.144	0.108	0.029	0.069	−0.762	−0.089	0.148	0.114	0.030	0.071
11	−0.730	−0.098	0.149	0.125	0.035	0.083	−0.732	−0.099	0.153	0.131	0.036	0.085
12	−0.466	−0.070	0.103	0.146	0.027	0.063	−0.466	−0.070	0.104	0.150	0.027	0.064
13	−0.091	−0.014	0.021	0.163	0.006	0.014	−0.091	−0.014	0.021	0.163	0.006	0.014
14	−0.243	−0.039	0.058	0.170	0.016	0.038	−0.242	−0.039	0.057	0.165	0.016	0.038
15	0.184	0.028	0.043	0.163	0.011	0.027	0.184	0.028	0.042	0.155	0.011	0.026
16	0.354	0.049	0.079	0.151	0.019	0.046	0.353	0.049	0.077	0.142	0.019	0.044
17	−0.027	−0.004	0.006	0.151	0.001	0.004	−0.027	−0.004	0.006	0.147	0.001	0.003
18	0.077	0.015	0.020	0.201	0.007	0.016	0.077	0.015	0.020	0.204	0.007	0.016
19	−0.235	−0.089	0.083	0.374	0.055	0.130	−0.236	−0.090	0.085	0.392	0.056	0.133
20	0.222	0.206	0.125	0.945	0.200	0.476	0.226	0.209	0.131	1.006	0.206	0.491

cutf: abbreviation for cut of point.

Table 6.4 Measures based on MM-estimates for a data set with three outliers (Cases 6, 7, and 8): mixture of MM with tangential leverage, Jacobian leverage and robust Jacobian leverage.

Measure	MM-estimate, tangential leverage							MM-estimate, Jacobian leverage						MM-estimate, robust Jacobian leverage					
	t_i	d_i	CD_i	p_{ii}	$DFFITS_i$	C_i	$DLEV_i$	t_i	d_i	CD_i	p_{ii}	$DFFITS_i$	C_i	t_i	d_i	CD_i	p_{ii}	$DFFITS_i$	C_i
cutf	3.000	3.000	1.000	0.375	0.775	2.000	-0.063	3.000	3.000	1.000	0.512	0.775	NA	3.000	3.000	1.000	0.395	0.775	2.000
1	0.200	0.006	0.021	0.032	0.001	0.002	0.027	0.196	0.006	NA	-0.004	NA	NA	0.203	0.006	0.029	0.062	0.001	0.003
2	1.657	0.000	0.207	0.047	0.000	0.000	-0.045	1.615	0.000	NA	-0.006	NA	NA	1.620	0.000	0.000	0.000	0.000	0.000
3	-1.020	0.000	0.150	0.065	0.000	0.000	0.056	-0.984	0.000	NA	-0.008	NA	NA	-1.051	0.000	0.221	0.133	0.000	0.000
4	-0.411	-0.083	0.070	0.088	0.025	0.059	0.075	-0.392	-0.079	NA	-0.011	NA	NA	-0.429	-0.087	0.106	0.185	0.036	0.085
5	1.271	0.000	0.247	0.113	0.000	0.000	0.096	1.196	0.000	NA	-0.014	NA	NA	1.345	0.000	0.386	0.247	0.000	0.000
6	*11.776*	0.000	*2.517*	*0.137*	0.000	0.000	*-0.121*	*10.961*	0.000	NA	-0.015	NA	NA	*11.044*	0.000	0.000	0.000	0.000	0.000
7	*10.606*	0.000	*2.400*	*0.154*	0.000	0.000	*-0.133*	*9.819*	0.000	NA	-0.011	NA	NA	*9.875*	0.000	0.000	0.000	0.000	0.000
8	*10.046*	0.000	*2.311*	*0.159*	0.000	0.000	*-0.137*	*9.336*	0.000	0.149	0.001	0.000	0.000	*9.332*	0.000	0.000	0.000	0.000	0.000
9	-0.915	0.000	0.207	0.154	0.000	0.000	0.105	-0.863	0.000	0.079	0.025	0.000	0.000	-0.976	0.000	0.315	0.312	0.000	0.000
10	0.043	-0.002	0.010	0.148	0.001	0.002	0.071	0.041	-0.002	0.006	0.064	0.000	0.001	0.045	-0.002	0.013	0.251	0.001	0.002
11	-0.360	-0.074	0.082	0.154	0.029	0.069	0.035	-0.353	-0.073	0.069	0.115	0.025	0.060	-0.367	-0.075	0.095	0.203	0.033	0.079
12	-0.023	-0.006	0.006	0.179	0.003	0.006	0.009	-0.023	-0.006	0.005	0.163	0.002	0.006	-0.023	-0.006	0.006	0.192	0.003	0.006
13	0.672	0.143	0.178	0.212	0.066	0.157	0.002	0.666	0.142	0.168	0.190	0.062	0.149	0.673	0.144	0.180	0.215	0.066	0.158
14	-0.486	-0.115	0.134	0.229	0.055	0.131	0.010	-0.477	-0.113	0.118	0.184	0.049	0.118	-0.489	-0.116	0.139	0.244	0.057	0.135
15	0.549	0.132	0.147	0.215	0.061	0.145	0.018	0.536	0.129	0.123	0.157	0.052	0.124	0.556	0.133	0.158	0.243	0.065	0.155
16	0.832	0.001	0.206	0.183	0.000	0.001	0.019	0.814	0.001	0.172	0.133	0.000	0.001	0.841	0.001	0.222	0.210	0.000	0.001
17	-0.639	-0.120	0.153	0.172	0.050	0.119	0.010	-0.630	-0.119	0.135	0.137	0.045	0.106	-0.643	-0.121	0.160	0.186	0.052	0.123
18	-0.199	-0.041	0.055	0.225	0.020	0.046	0.000	-0.196	-0.040	0.049	0.185	0.018	0.042	-0.199	-0.041	0.055	0.225	0.020	0.047
19	-1.044	-0.001	0.377	*0.390*	0.000	0.001	0.002	-1.005	-0.001	0.312	0.288	0.000	0.001	-1.046	-0.001	0.380	*0.395*	0.000	0.001
20	*1.008*	*1.086*	*0.514*	*0.779*	*0.958*	*2.281*	*0.022*	*0.913*	*0.983*	*0.358*	*0.460*	*0.736*	*1.753*	*1.029*	*1.108*	*0.549*	*0.853*	*1.002*	*2.386*

Empty cells are shown where values cannot be calculated because of super leverage or the singularity matrix.
cutf: abbreviation for cut of point. NA: values can not be computed.

Equation 6.17 cannot be computed because the inverse of the matrix $(\hat{V}^T\hat{V} - [\hat{\mathbf{r}}] \otimes [\hat{W}])$ does not exist,

$$(\hat{V}^T\hat{V} - [\hat{\mathbf{r}}] \otimes [\hat{W}])$$

$$= \left\{ \begin{array}{ccc} 5.152325 & -66.54006 & 136417.7 \\ -66.540065 & 723.67547 & -2333671.6 \\ 36417.707216 & -2333671.57659 & 4479954768.1 \end{array} \right\}$$

With eigenvalues, this equals

$$(4.479956e + 009, 1.039786e + 000, -4.920083e + 002)$$

This means that the nonrobust Jacobian leverage cannot be computed. The residuals computed from the MM-estimate will be large and create a rounding error in the computation. This shows that, instead of using residual $[\mathbf{r}]$ in Equation 6.9, it is better to use the robust form $\psi(\mathbf{r})$, as in Equation 6.17.

The results of seven outlier measures for Case C are presented in Tables 6.5 and 6.6. The presence of seven high leverage points makes it harder for almost all outlier detection methods to detect high leverage points correctly. In this situation, most detection measures based on OLS estimates fail to identify even a single high leverage point because of masking effects. It can be seen from Table 6.6 (nonrobust Jacobian leverage) that again t_i-MM and CD_i-MM do a credible job. Both measures can identify the six high leverage points correctly. However, in some situations the nonrobust Jacobian leverage has computational problems. The results in Table 6.6, which are based on the robust Jacobian leverage, show that only t_i-MM and CD_i-MM correctly identify the outliers, as for Cases A and B. t_i based on robust Jacobian leverage correctly identifies the outliers.

6.6 Numerical Example

In this section, real data – the lake data – are presented (Stromberg 1993) and used to compare the preceding methods. The data set was collected from 29 lakes in Florida by the United States Environmental Protection Agency (1978). Stromberg (1993) has identified observations 10 and 23 as outliers.

The data present the relationship between the mean annual total nitrogen concentration (TN; the response variable), and the average influence nitrogen concentration (NIN) and water retention time (TW) as predictors. The model associated with the data is

$$TN_i = \frac{NIN_i}{1 + \delta TW_i^\beta} + \epsilon_i, i = 1, \dots, 29 \tag{6.23}$$

with unknown parameter vector $\theta = (\delta, \beta)$. The results for measures based on tangent plane leverages are shown in Table 6.7. For the reason discussed in

Table 6.5 Measures based on OLS estimates for a data set with six outliers (last six observations in Cases 6, 7, and 8): mixture of OLS with tangential leverage and Jacobian leverage.

Measure	OLS estimate, tangential leverage						OLS estimate, Jacobian leverage					
	t_i	d_i	CD_i	p_{ii}	$DFFITS_i$	C_i	t_i	d_i	CD_i	p_{ii}	$DFFITS_i$	C_i
cutf	3.000	3.000	1.000	0.256	0.679	2.000	3.000	3.000	1.000	0.243	0.679	2.000
1	−0.673	−0.021	0.074	0.036	0.004	0.011	−0.671	−0.020	0.068	0.030	0.004	0.010
2	0.209	0.007	0.025	0.043	0.002	0.004	0.209	0.007	0.023	0.036	0.001	0.004
3	−1.563	−0.070	0.201	0.050	0.016	0.043	−1.557	−0.070	0.185	0.042	0.014	0.040
4	−1.192	−0.060	0.165	0.057	0.014	0.040	−1.188	−0.060	0.151	0.049	0.013	0.037
5	−0.114	−0.006	0.017	0.065	0.002	0.004	−0.113	−0.006	0.015	0.055	0.001	0.004
6	−0.210	−0.013	0.033	0.073	0.003	0.010	−0.209	−0.013	0.030	0.062	0.003	0.009
7	−0.878	−0.060	0.143	0.080	0.017	0.047	−0.873	−0.060	0.132	0.068	0.016	0.044
8	−1.058	−0.079	0.179	0.086	0.023	0.064	−1.052	−0.078	0.165	0.074	0.021	0.059
9	−0.874	−0.068	0.151	0.090	0.020	0.056	−0.869	−0.068	0.140	0.078	0.019	0.053
10	0.052	0.004	0.009	0.092	0.001	0.003	0.052	0.004	0.009	0.081	0.001	0.003
11	0.163	0.013	0.029	0.091	0.004	0.011	0.163	0.013	0.027	0.081	0.004	0.010
12	0.741	0.060	0.127	0.089	0.018	0.049	0.738	0.059	0.121	0.081	0.017	0.047
13	1.480	0.123	0.250	0.086	0.036	0.100	1.476	0.123	0.242	0.080	0.035	0.097
14	0.973	0.080	0.163	0.085	0.023	0.064	0.971	0.080	0.160	0.082	0.023	0.063
15	1.637	0.150	0.281	0.089	0.045	0.124	1.636	0.150	0.281	0.088	0.045	0.124
16	1.637	0.177	0.304	0.103	0.057	0.157	1.638	0.177	0.305	0.104	0.057	0.158
17	0.276	0.036	0.059	0.135	0.013	0.037	0.276	0.036	0.058	0.135	0.013	0.037

(Continued)

Table 6.5 (Continued)

| Measure | OLS estimate, tangential leverage | | | | | | OLS estimate, Jacobian leverage | | | | | |
| | t_i | d_i | CD_i | p_{ii} | $DFFITS_i$ | C_i | t_i | d_i | CD_i | p_{ii} | $DFFITS_i$ | C_i |
cutf	3.000	3.000	1.000	0.256	0.679	2.000	3.000	3.000	1.000	0.243	0.679	2.000
18	−0.161	−0.030	0.041	0.195	0.013	0.036	−0.160	−0.030	0.040	0.189	0.013	0.036
19	−1.729	−0.512	0.545	0.298	0.279	0.774	−1.717	−0.509	0.524	0.279	0.271	0.750
20	−1.930	−0.891	0.763	0.469	0.610	1.690	−1.899	−0.877	0.713	0.423	0.580	1.605
21	0.625	0.120	0.160	0.197	0.053	0.147	0.624	0.120	0.159	0.195	0.053	0.147
22	−0.056	−0.011	0.014	0.200	0.005	0.014	−0.056	−0.011	0.014	0.200	0.005	0.014
23	−1.201	−0.261	0.322	0.216	0.121	0.336	−1.200	−0.261	0.321	0.215	0.121	0.335
24	−0.337	−0.071	0.090	0.216	0.033	0.092	−0.337	−0.071	0.090	0.215	0.033	0.091
25	1.339	0.265	0.343	0.197	0.118	0.326	1.337	0.265	0.341	0.195	0.117	0.324
26	−0.099	−0.023	0.028	0.241	0.011	0.031	−0.099	−0.023	0.028	0.238	0.011	0.031

cutf: abbreviation for cut of point.

Table 6.6 Measures based on MM-estimates for a data set with six outliers (last six observations in Cases 6, 7, and 8): mixture of MM with tngential leverage, Jacobian leverage and robust Jacobian leverage.

Measure	MM-estimate, tangential leverage								MM-estimate, Jacobian leverage						MM-estimate, robust Jacobian leverage				
	t_i	d_i	CD_i	p_{ii}	$DFFITS_i$	C_i	$DLEV_i$	t_i	d_i	CD_i	p_{ii}	$DFFITS_i$	C_i	t_i	d_i	CD_i	p_{ii}	$DFFITS_i$	C_i
cutf	3.000	3.000	1.000	0.261	0.679	2.000	−0.075	3.000	3.000	1.000	0.231	0.679	2.000	3.000	3.000	1.000	0.459	0.679	2.000
1	0.157	0.005	0.012	0.017	0.001	0.002	0.020	0.156	0.005	0.010	0.013	0.001	0.002	0.158	0.005	0.018	0.039	0.001	0.003
2	1.382	0.000	0.129	0.026	0.000	0.000	−0.025	1.378	0.000	0.110	0.019	0.000	0.000	1.364	0.000	0.000	0.000	0.000	0.000
3	−0.872	0.000	0.098	0.038	0.000	0.000	0.036	−0.868	0.000	0.085	0.029	0.000	0.000	−0.889	0.000	0.143	0.078	0.000	0.000
4	−0.361	−0.035	0.048	0.054	0.008	0.023	0.044	−0.359	−0.035	0.042	0.041	0.007	0.020	−0.370	−0.036	0.069	0.105	0.011	0.032
5	1.043	0.000	0.163	0.073	0.000	0.000	0.050	1.036	0.000	0.144	0.058	0.000	0.000	1.073	0.000	0.227	0.134	0.000	0.000
6	0.846	0.000	0.151	0.096	0.000	0.000	0.051	0.839	0.000	0.136	0.078	0.000	0.000	0.871	0.000	0.202	0.161	0.000	0.000
7	−0.195	−0.032	0.039	0.119	0.011	0.030	0.045	−0.194	−0.031	0.036	0.101	0.010	0.028	−0.200	−0.033	0.049	0.178	0.013	0.037
8	−0.684	−0.113	0.146	0.137	0.042	0.116	0.032	−0.680	−0.112	0.138	0.123	0.039	0.109	−0.697	−0.115	0.171	0.180	0.048	0.132
9	−0.780	−0.122	0.173	0.147	0.047	0.129	0.015	−0.778	−0.121	0.168	0.140	0.046	0.126	−0.787	−0.123	0.186	0.168	0.050	0.138
10	0.042	0.005	0.009	0.147	0.002	0.006	0.005	0.042	0.005	0.009	0.149	0.002	0.006	0.042	0.005	0.010	0.153	0.002	0.006
11	−0.287	−0.042	0.062	0.139	0.016	0.043	0.010	−0.288	−0.042	0.064	0.148	0.016	0.044	−0.289	−0.042	0.065	0.152	0.016	0.045
12	0.007	0.001	0.002	0.133	0.000	0.001	0.031	0.007	0.001	0.002	0.141	0.000	0.001	0.007	0.001	0.002	0.174	0.001	0.001
13	0.588	0.122	0.125	0.136	0.045	0.124	0.055	0.587	0.121	0.123	0.133	0.044	0.123	0.607	0.125	0.161	0.211	0.056	0.154
14	−0.362	−0.081	0.080	0.147	0.031	0.086	0.063	−0.359	−0.081	0.074	0.128	0.029	0.080	−0.376	−0.084	0.105	0.236	0.039	0.109
15	0.492	0.108	0.114	0.162	0.043	0.120	0.046	0.485	0.106	0.099	0.126	0.038	0.106	0.506	0.111	0.139	0.227	0.051	0.142

(Continued)

Table 6.6 (Continued)

		MM-estimate, tangential leverage					MM-estimate, Jacobian leverage								MM-estimate, robust Jacobian leverage				
Measure	t_i	d_i	CD_i	p_{ii}	$DFFITS_i$	C_i	d_i	$DLEV_i$	t_i	CD_i	p_{ii}	$DFFITS_i$	C_i	t_i	d_i	CD_i	p_{ii}	$DFFITS_i$	C_i
cutf	3.000	3.000	1.000	0.261	0.679	2.000	3.000	-0.075	3.000	1.000	0.231	0.679	2.000	3.000	3.000	1.000	0.459	0.679	2.000
16	0.731	0.140	0.175	0.171	0.058	0.161	0.138	0.016	0.717	0.146	0.125	0.050	0.138	0.738	0.142	0.188	0.194	0.062	0.171
17	-0.530	-0.092	0.126	0.170	0.038	0.105	-0.090	0.005	-0.519	0.105	0.124	0.032	0.089	-0.531	-0.092	0.129	0.177	0.039	0.107
18	-0.168	-0.035	0.039	0.159	0.014	0.039	-0.035	0.045	-0.165	0.033	0.120	0.012	0.034	-0.172	-0.036	0.047	0.223	0.017	0.046
19	-0.826	-0.331	0.180	0.143	0.125	0.347	-0.327	0.155	-0.816	0.161	0.117	0.113	0.313	-0.911	-0.365	0.328	0.390	0.207	0.572
20	0.646	0.525	0.134	0.129	0.189	0.522	0.522	0.332	0.642	0.125	0.114	0.177	0.491	0.817	0.665	0.423	0.806	0.472	1.306
21	37.888	0.000	9.382	0.184	0.000	0.000	0.000	-0.155	37.530	8.712	0.162	0.000	0.000	34.821	0.000	0.000	0.000	0.000	0.000
22	37.994	0.000	9.435	0.185	0.000	0.000	0.000	-0.156	37.626	8.749	0.162	0.000	0.000	34.902	0.000	0.000	0.000	0.000	0.000
23	36.972	0.000	9.193	0.185	0.000	0.000	0.000	-0.156	36.612	8.519	0.162	0.000	0.000	33.957	0.000	0.000	0.000	0.000	0.000
24	38.092	0.000	9.471	0.185	0.000	0.000	0.000	-0.156	37.721	8.777	0.162	0.000	0.000	34.986	0.000	0.000	0.000	0.000	0.000
25	38.821	0.000	9.613	0.184	0.000	0.000	0.000	-0.155	38.454	8.927	0.162	0.000	0.000	35.678	0.000	0.000	0.000	0.000	0.000
26	38.844	0.000	9.669	0.186	0.000	0.000	0.000	-0.157	38.462	8.956	0.163	0.000	0.000	35.670	0.000	0.000	0.000	0.000	0.000

Empty cells are shown where values cannot be calculated because of super leverage or the singularity matrix.
cutf: abbreviation for cut of point.

Table 6.7 Outlier measures for lakes data, computed using the hat matrix and Jacobian leverage, and the MM-estimator.

Cut-off point index	t_i 3.0	CD_i 1.0	p_i 0.079	JL d_i 3.0	JL CD_i 1.0	JL p_i 0.187	$DLEV_i$ −0.030
1	−1.423	0.418	0.173	−1.440	0.456	0.201	0.020
2	1.348	0.142	0.022	1.552	0.653	0.354	0.240
3	0.035	0.002	0.008	0.035	0.003	0.016	0.008
4	−0.451	0.028	0.008	−0.475	0.117	0.120	0.100
5	0.999	0.099	0.020	1.023	0.190	0.069	0.045
6	−0.930	0.070	0.011	−0.947	0.145	0.047	0.034
7	−0.932	0.162	0.061	−0.937	0.177	0.071	0.009
8	−0.632	0.039	0.008	−0.666	0.162	0.118	0.098
9	1.552	0.166	0.023	1.535	0.000	0.000	−0.022
10	−9.709	5.182	0.570	−7.749	0.000	0.000	−0.363
11	−1.075	0.233	0.094	−1.081	0.249	0.106	0.010
12	0.275	0.011	0.003	0.276	0.016	0.006	0.003
13	−0.422	0.027	0.008	−0.424	0.041	0.018	0.010
14	0.943	0.109	0.027	0.947	0.124	0.034	0.007
15	0.347	0.025	0.010	0.348	0.029	0.014	0.003
16	1.246	0.356	0.163	1.258	0.383	0.186	0.016
17	0.757	0.046	0.007	0.765	0.091	0.028	0.020
18	−0.012	0.000	0.003	−0.012	0.001	0.004	0.001
19	−0.041	0.001	0.002	−0.041	0.002	0.003	0.001
20	−0.049	0.008	0.051	−0.049	0.009	0.061	0.009
21	0.785	0.220	0.157	0.797	0.246	0.191	0.024
22	−0.556	0.072	0.033	−0.658	0.310	0.445	0.276
23	−29.263	30.200	2.130	−16.540	0.000	0.000	−0.681
24	1.586	0.085	0.006	1.581	0.000	0.000	−0.006
25	−0.320	0.026	0.013	−0.320	0.028	0.015	0.002
26	−0.660	0.042	0.008	−0.665	0.069	0.022	0.013
27	−0.898	0.115	0.033	−0.918	0.182	0.079	0.042
28	0.173	0.039	0.100	0.174	0.042	0.118	0.015
29	−0.261	0.010	0.003	−0.263	0.022	0.014	0.011

JL, computation performed by Jacobian leverage. For underlined values the measure is greater than the cut-off point. For *DLEV* the value is lower than than negative cut-off point.
Source: Detection in Nonlinear Regression, First A. Hossein Riazoshams, Second B. Midi Habshah, Jr., Third C. Mohamad Bakri Adam, World academy of science 2011. Adapted from A. Hossein Riazoshams (2011), B. Midi Habshah, Jr (2011), C. Mohamad Bakri Adam (2011).

Section 6.4.9, only measures that are not based on deleting points and that use the robust MM-estimate are displayed. The Cook distance, studentized residuals, $DLEV$, Hadi potential, and studentized residuals when combined with Jacobian leverage are successful in isolating outliers. Figure 8.9 displays the tangential plane leverage and robust Jacobian leverage, and their differences. As noted in Section 6.4.8, the tangential plan leverage is larger than the robust Jacobian leverage for outlier data, so their differences will be large and negative. As can be seen from Figure 8.9, the large negative value (not positive) for data points 10 and 23 clearly represents the downgraded point in the MM-estimation procedure.

6.7   Variance Heteroscedasticity

This section extends several measures of outlier detection used in linear and nonlinear regression to identify the outliers in nonlinear regression that have a heteroscedastic error when the variance of error follows an unknown parameteric function of the predictors. A new concept of variance outliers is defined and their effect on inference studied in Chapter 4. The proposed methods use robust estimation of nonlinear regression parameters and variance function parameters, and modify several statistical measures of influence to enable them to be used in the variance heteroscedastic function.

As discussed in previous sections, the standardized residual and Cook distance, when combined with robust estimators, can identify single or multiple outliers. For this reason, in this section these two statistics will be extended to identification of outliers in nonlinear regression when there are heteroscedastic variance errors. The intention is to extend the standardized residual and Cook distance to nonlinear regression. The formulas will then be written for generalized least squares estimators and finally derived for robust MM-estimators. Furthermore, other statistical measures will also be rewritten in a generalized model.

Heterogeneity of variance is a problematic situation. Riazoshams and Midi (2016) discussed two situations: data point outliers and variance outliers. In the first, the data points were outliers while their variance was not (Figure 6.2b) and in the second, a single outlier point created a variance innovation outlier (IO), which is an outlier that affects the next consecutive point. This occurs in the nonreplicated case shown in Figure 6.3.

This section also presents other cases. The data points might not be outliers but due to small variations their variances might be extremely small. As shown in Figure 6.2c, data at $x_i = 51$ are aggregated, so variance at the point is an outlier in the x direction. Outlier data points at a single x_i, although repeated, include some outliers that create outliers in variance as well, but at such a point there will be too many outliers at one cross-section, so even robust estimates of variance parameters break down (see Figure 6.4). These examples show the unexpected effect of outliers on data and variance, and thus motivate us not

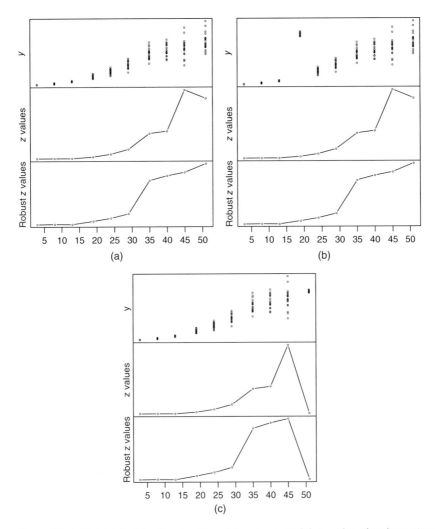

Figure 6.2 Artificial data: (a) without outliers, (b) contaminated data with outlier data points but variances are not outliers, (c) contaminated data with extremely small variance but data points are not outliers. *Source*: Robust Multistage Estimator in Nonlinear Regression with Heteroscedastic Errors Hossein Riazoshams & Habshah BT. Midi, Communications in Statistics – Simulation and Computation. Adapted from Author's Hossein Riazoshams & Habshah BT. Midi.

only to identify the outlier points, but also to try to define and identify variance outliers. It is important to consider the latter as they intertwine and interact with estimates.

Another effect of outliers is masking and swamping. Masking happens when outliers are hidden, looking as though they are good points, and swamping happens when the presence of outliers suggests that good data is bad. Furthermore, masking and swamping effects can be seen in the

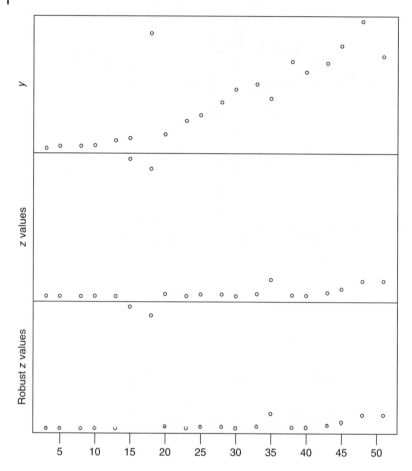

Figure 6.3 Artificial data, nonreplicated data.

presence of heterogeneity of error variance; that is, outliers might suggest that heterogeneous variance is homogeneous – masking heterogeneity – or suggest that homogeneous variance is heterogeneous – swamping. To resolve such problems, both parameters of the nonlinear regression model θ and variance model τ should be estimated with robust estimators. Furthermore, the statistics for identifying outliers should be modified appropriately and mixed with robust estimators.

6.7.1 Heteroscedastic Variance Studentized Residual

As shown in previous sections (and in Riazoshams et al. (2011)), standardized residuals can identify outliers when the parameter estimates and standard error

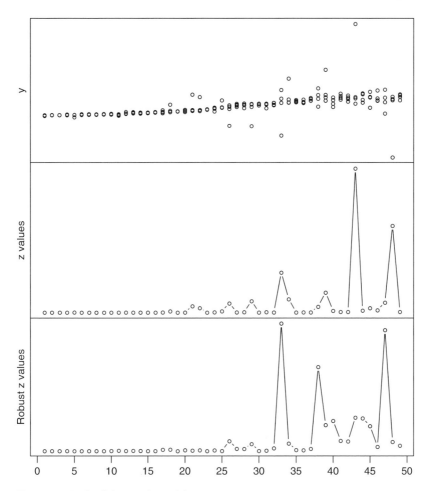

Figure 6.4 Artificial data, replicated data.

estimate $\hat{\sigma}$ are replaced by robust MM-estimators. In this section, our purpose is to derive the standardized residual for generalized least squares estimators for the heterogeneous variance case and robust estimators. First, to derive the formulas for generalized least squares estimators (Equation 2.17), recall that

$$\hat{\theta}_{GLS} = \arg\min_{\theta} [\mathbf{Y} - \boldsymbol{\eta}(\theta)]^T V^{-1} [\mathbf{Y} - \boldsymbol{\eta}(\theta)] \tag{6.24}$$

Assume the variance of error follows the heteroscedastic form (Equation 4.4)

$$Var(\varepsilon) = \mathrm{diag}[G(x_i; \tau)] \tag{6.25}$$

or a form with overall constant variance term σ^2 (Equation 4.8), that is,

$$Var(\varepsilon) = \sigma^2 \mathrm{diag}[g(x_i, \lambda)] \tag{6.26}$$

which can be represented by $Cov(\varepsilon) = \sigma^2 V(\lambda)$. Then, using the Cholesky decomposition of $V = U^T U$ and transforming the nonlinear regression model by $R = (U^T)^{-1}$, both of which depend on λ, we have

$$\mathbf{Z} = \kappa(\theta) + \xi \tag{6.27}$$

where $\mathbf{Z} = R\mathbf{Y}$ are transformed response values, $\kappa = R\eta$ is the transformed nonlinear regression, and $\xi = R\varepsilon$ is the transformed error. This is a nonlinear regression model with constant variance $Cov(\xi) = \sigma^2 I$, so the linear approximation (6.2) can be derived for the transformed model 6.27. The gradient of the transformed nonlinear model is

$$\dot{\kappa}(\theta) = \frac{\partial \kappa(\theta)}{\partial \theta} = R\dot{\eta}(\theta) \tag{6.28}$$

Note that the matrix R depends on variance model parameter λ and will be substituted by its estimates. Using this transformation, it is easy to show that generalized least squares is similar to the ordinary least squares estimate (see Seber and Wild (2003)). The sum of the squared errors for the transformed model 6.27 is

$$S(\theta) = \|\mathbf{Z} - \kappa(\theta)\|^2$$
$$\approx \|\mathbf{Z} - \kappa(\theta_0) - \dot{\kappa}(\theta_0)(\theta - \theta_0)\|^2$$

which leads us to the minimization Equation 6.24. Thus the generalized normal equations of (6.3) are

$$[\dot{\kappa}^T(\theta_0)\dot{\kappa}(\theta_0)]\beta = \dot{\kappa}^T(\theta_0)\xi$$

The hat matrix for generalized least squares, denoted by $\mathbb{H}$, can be written as:

$$\mathbb{H} = \dot{\kappa}(\theta_0)[\dot{\kappa}^T(\theta_0)\dot{\kappa}(\theta_0)]^{-1}\dot{\kappa}^T(\theta_0)$$
$$= R\dot{\eta}(\theta_0)[\dot{\eta}^T(\theta_0)V^{-1}\dot{\eta}(\theta_0)]^{-1}\dot{\eta}^T(\theta_0)R^T$$

in which $\dot{\kappa}(\theta_0)$ is gradient (6.28) calculated at true value θ_0. In practice, the parameter's true value is unknown and the generalized least squares estimator, and consequently $\dot{\kappa}(\hat{\theta}_{GLS})$, is estimated using $\hat{\theta}_{GLS}$. Similarly, the residuals can be computed by:

$$\mathbf{r} = \mathbf{Z} - \kappa(\hat{\theta}_{GLS})$$
$$= (I - \mathbb{H})\xi.$$

Therefore, the standardized residuals for generalized least squares estimators are

$$t_{GLi} = \frac{\mathbf{Z} - \kappa(\hat{\theta}_{GLS})}{\hat{\sigma}\sqrt{(1 - \mathbb{H}_{ii})}}$$
$$= \frac{R[y_i - f(x_i; \hat{\theta}_{GLS})]}{\hat{\sigma}_{GLS}\sqrt{1 - \mathbb{H}_{ii}}}. \tag{6.29}$$

where $\mathbb{H}_{ii}$ is the ith diagonal element of G. The estimate of constant variance σ can be calculated from the final sum of squared errors

$$(n-p)\hat{\sigma}_{GLS} = S(\hat{\theta}_{GLS})$$

Analogously, other measures can be computed using the transformed model in Equation 6.27. For example, the Cook distance (6.20) can be rewritten as

$$CD_{GLi}(V, p\sigma^2) = \frac{t_{GLi}^2}{p} \frac{\mathbb{H}_{ii}}{1 - \mathbb{H}_{ii}}$$

To formulate this more systematically for computational purposes, the non-linear regression model and residuals must be transformed. The transformed values can then be substituted in the outlier detection measures.

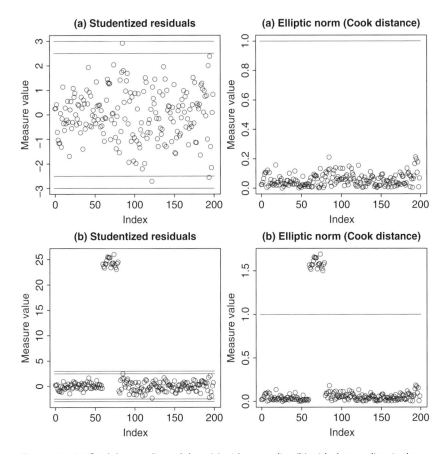

Figure 6.5 Artificial data, replicated data: (a) without outlier, (b) with data outliers in the middle but their variance is not outlier.

6.7.2 Simulation Study, Heteroscedastic Variance

As seen in earlier sections of this chapter, only the studentized residual, Cook distance, and *DLEV* are capable of identifying outliers correctly in most cases. *DLEV* for the generalized nonlinear model has not been developed yet, so for the heteroscedastic error case only the studentized residual and Cook distance will be reported. Consider the simulated cases in Figures 6.5 and 6.6. The figures show the calculated studentized residual and Cook distance for the simulated data. Note that the heteroscedasticity of variance is removed from the curves by the transformed model. Figure 6.6a,b display the measures for the simulated data in which the data at the last point are squared at the end; that is, the variance outlier but not the data point outlier case. The studentized residuals display a small range of variation at the end, which might help us to identify variance outliers.

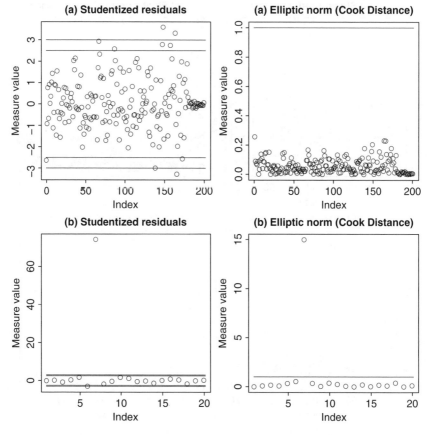

Figure 6.6 Artificial replicated and nonreplicated data: (a) replicated data with small variance outlier at the end of the data series, (b) nonreplicated data with a single outlier in the middle.

6.8   Conclusion

For identifying outliers, the best combination is a t_i, CD_i-MM-based estimate, with tangential leverage and $DLEV$ (see Riazoshams et al. (2011)).

In some cases, the nonrobust Jacobian leverage $\hat{J}$ gives a similar result. However, occasionally it creates computational problems due to large residual values. As a result, Jacobian leverage is not usually incorporated in the formulation of an outlier detection measure. On the other hand, Jacobian leverage gives smaller values to outliers and does not involve any computation problems.

In this chapter, we derived a robust Jacobian leverage and used it to formulate a reliable alternative outlier detection measure. The advantage of this new measure is that its formulation does not depend on the other six measures discussed earlier. Additionally it is computationally simple. It is a very good way to correctly identify outliers in the data.

We also incorporated the nonrobust and robust Jacobian leverages in six outlier detection measures. The results reveal that only t_i-MM-based estimates, when combined with robust Jacobian leverage, can successfully identify the outliers. The t_i-MM-based and CD_i-MM-based estimates combined with nonrobust Jacobian leverage also can identify outliers. However, the nonrobust Jacobian leverage has computational problems.

For the heteroscedastic variance cases, the elliptic norm (Cook distance) and studentized residuals have been developed, and these do a good job in detecting atypical points.

Part Two

Computations

7

Optimization

For all the methods discussed in the previous chapters we saw numerous examples that required the numerical minimization of a nonlinear function. For example, the most efficient method, the maximum likelihood estimate equation, requires the likelihood function to be minimized (2.7). However, in general form there is no explicit answer to this problem. Unlike linear regression, in a general nonlinear regression model there is no explicit formula for parameter estimates in any of the discussed methods. This fact creates a vital need for numerical optimization techniques. The problems that may arise in optimization are the convergence of algorithms and sensitivity to an initial value. In addition, in robust statistics the effect of outliers on rounding errors is an issue. Outliers have a large effect on numerical computation and on the computation of the objective function, gradients and Hessian matrices.

This chapter discusses the optimization methods applied to the robust nonlinear regression estimators used in the book, as well as situations that involve with outliers and their rounding error effects. We have observed that outliers not only have a null effect during iteration procedures, but also affect the initial values, and this has an effect on the overall convergence of iteration. Accordingly, we need iterative optimization methods that are robust against outlier rounding error effects, are less sensitive to initial values, and that have been developed for robust nonlinear regression.

7.1   Optimization Overview

For minimization of least squares estimates, the modified Gauss–Newton procedure developed by Bates and Watts (2007) is used in the `nlr` package for OLS estimates. For minimization of robust objective functions, we need more specialized methods, and these are discussed in this chapter. These optimization methods are applied for computing the estimators discussed in the book, and are also used in the `nlr` package.

Robust Nonlinear Regression: with Applications using R, First Edition.
Hossein Riazoshams, Habshah Midi, and Gebrenegus Ghilagaber.
© 2019 John Wiley & Sons Ltd. Published 2019 by John Wiley & Sons Ltd.
Companion website: www.wiley.com/go/riazoshams/robustnonlinearregression

This chapter covers specialized numerical techniques for the optimization of robust objective functions. Convergence, the effect of outliers and initial values in numerical iterations, stopping rules, and resolution of singularity problems are discussed. The techniques use first derivatives (gradient matrix) and second derivatives (Hessian matrix) of the parameters, so we will focus on functions that have first and second derivatives. This assumption does not limit the theory, but only its computational aspects. Note that the `nlr` package (Chapter 8) provided alongside this book is based on the existence of derivatives, but includes derivative-free methods as well, although are less efficient and exhibit slower convergence.

The gradient and Hessian matrices values are affected by outliers, and consequently exhibit rounding error effects and singularity problems. To solve this problem, three techniques are combined: the Newton, steepest descent, and Levenberg–Marquardt methods. Together these algorithms provide fast convergence in clean data and good accuracy in the presence of outliers, but they still need more study to improve their accuracy. Fortunately, our experience in different situations suggests that they provide sufficiently reliable results, although they are unstable in some situations.

It is important to note that nonlinear regression models are more general than linear regression models so the choice of functions is unlimited. It is therefore impossible to find a universal technique to resolve all nonlinear regression models for any kind of data set. In practice the `nlr` package provides a few choices from which to select an appropriate iterative method.

7.2 Iterative Methods

Suppose we want to minimize a general function $y = w(\theta)$ (the objective function), defined in the p-dimensional domain $\theta \in \Re^p$. The p-vector of the gradient, denoted by ∇, is defined as

$$\nabla = \frac{\partial w}{\partial \theta} = \left[\frac{\partial w}{\partial \theta_1}, \ldots, \frac{\partial w}{\partial \theta_p} \right]^T$$

and the $p \times p$ symmetric Hessian matrix, denoted by H, is defined as

$$H = \frac{\partial^2 w}{\partial \theta \partial \theta^T} = \begin{bmatrix} \frac{\partial^2 w}{\partial \theta_1^2} & \frac{\partial^2 w}{\partial \theta_1 \partial \theta_2} & \cdots & \frac{\partial^2 w}{\partial \theta_1 \partial \theta_p} \\ \frac{\partial^2 w}{\partial \theta_2 \partial \theta_1} & \frac{\partial^2 w}{\partial \theta_2^2} & \cdots & \frac{\partial^2 w}{\partial \theta_2 \partial \theta_p} \\ \vdots & \vdots & \vdots & \vdots \\ \frac{\partial^2 w}{\partial \theta_p \partial \theta_1} & \frac{\partial^2 w}{\partial \theta_p \partial \theta_2} & \cdots & \frac{\partial^2 w}{\partial \theta_p^2} \end{bmatrix}$$

The domain θ, in most but not all cases in nonlinear regression, is a function model parameter.

When we have a sample of data points, for example $(x_i, w_i(\theta))$, $w_i(\theta) = w(x_i, \theta))$, $i = 1, \ldots, n$, similar to the nonlinear regression model $y_i = f(x_i; \theta) + \varepsilon_i$, a sample of n points for the gradient and Hessian matrices is available. The gradient is stored in a $n \times p$ matrix, for simplicity denoted using the same notation, and the Hessian in a $n \times p \times p$ three-dimensional array:

$$\nabla w(\theta) = \begin{bmatrix} \dfrac{\partial w_1}{\partial \theta_1} & \dfrac{\partial w_1}{\partial \theta_2} & \cdots & \dfrac{\partial w_1}{\partial \theta_p} \\[2ex] \dfrac{\partial w_2}{\partial \theta_1} & \dfrac{\partial w_2}{\partial \theta_2} & \cdots & \dfrac{\partial w_2}{\partial \theta_p} \\[2ex] \vdots & \vdots & \vdots & \vdots \\[2ex] \dfrac{\partial w_n}{\partial \theta_1} & \dfrac{\partial w_n}{\partial \theta_2} & \cdots & \dfrac{\partial w_n}{\partial \theta_p} \end{bmatrix}$$

$$H(\theta) = [w_{ijk}]_{n \times p \times p},$$

$$w_{ijk} = \frac{\partial^2 w_i(\theta)}{\partial \theta_j \partial \theta_k}, \quad i = 1, \ldots, n, \; j = 1, \ldots, p, \; k = 1, \ldots, p.$$

To minimize the objective function $w(\theta$, we use iteration. At each step the parameter θ is updated. If the step iteration is denoted by (a), at step $(a + 1)$ the parameter θ can be evaluated by increments δ using the line search method:

Line search method: $\qquad \theta^{(a+1)} = \theta^a + \delta^{(a)}$ $\qquad\qquad$ (7.1)

At each iteration, a new value of the objective function is computed at $w^{(a+1)} = w(\theta^{(a+1)})$, and the decrease in its value is checked.

The Newton method computes the increments δ as follows:

Newton method: $\qquad \delta^{(a)} = -H^{-1(a)} \nabla^{(a)}$

where $H^{-1(a)}$ is the inverse of the Hessian matrix and $\nabla^{(a)}$ is calculated by the gradient matrix at step (a). The Newton method does not reduce the function values. For this reason, the steepest descent method computes the increment by adding the coefficient to provide a decrease in the objective function

Steepest descent method: $\qquad \delta^{(a)} = -\alpha^{(a)} H^{-1(a)} \nabla^{(a)}$

The coefficient $\alpha^{(a)}$ is called the step size and increases by a large factor, say 10, if the function decreases. This provides faster convergence.

In the presence of outliers, the Hessian matrix computed at outlying points can be far from other cells and, due to rounding errors, it will be singular, although mathematically the Hessian is nonsingular. To solve this nonsingularity problem and to ensure that the search direction is a descent direction, the Levenberg–Marquardt technique is applied. If a constant value is added to the diagonal of the Hessian matrix, as follows, the minimization objective function does not change (Chong and Zak,1996):

Levenberg–Marquardt method: $\qquad \delta^{(a)} = -(H + \mu_a I)^{-1(a)} \nabla^{(a)}$

where $\mu_a \geq 0$ is sufficient large. There are some choices for this value, for example the smallest eigenvalue of the Hessian matrix:

$$\delta^{(a)} = -(H + \lambda I)^{-1(a)}\nabla^{(a)}$$

$$\lambda = \min(\lambda_1, \ldots, \lambda_n) \tag{7.2}$$

where λ_i are eigenvalues of H. It can been seen that this change gives the descent direction in some cases. When the curvature of the objective function is very different to the direction of the parameter, and the outliers are large, the cells of the Hessian matrix are far apart and cause a rounding error computation problem. For this reason another choice is to use all eigenvalues to move all diagonals, but then it is necesary to add a small value to create a matrix with positive eigenvalues. In other words, a small amount is added to change negative values into positive ones:

$$\delta^{(a)} = -(H + diag[v_i]_{(a)}I)^{-1(a)}\nabla^{(a)}$$

$$v_i = \begin{cases} \lambda_i & \lambda > 0 \\ c - \lambda_i & \lambda < 0 \end{cases} \tag{7.3}$$

The nlr package uses $c = 10^{-8}$. We have found that none of these techniques is successful in all situations, so to achieve overall convergence the nlr package must switch between methods if one fails.

7.3 Wolfe Condition

The Wolfe condition in nlr is an alternative for when the above algorithms do not converge. It is an inexact line search method that stipulates $\theta^{(a)}$ in the line search method (Equation 7.1) – it must give a sufficient decrease in the objective function and a suitable curvature area should be chosen to ensure a significant reduction. These two conditions are known as the Wolfe conditions. By writing the line search method (Equation 7.1) for stepsize α_a and direction p_a as

$$\theta^{(a+1)} = \theta^a + \alpha_a p_a$$

the decrease, known as the Armijo condition, is written as the inequality

$$w(\theta^{(a)} + \alpha p_a) \leq w(\theta^{(a)}) + c_1 \alpha \nabla^T w_a p_a$$

for some constant $c_1 \in (0,1)$. The curvature condition is written as the inequality

$$\nabla^T w(\theta^{(a)} + \alpha_a p_a)p_a \geq c_2 \nabla^T w^{(a)} p_a$$

for some constant $c_2 \in (0,1), 0 < c_1 < c_2 < 1$.

As the descent direction method, the nlr package uses the Levenberg–Marquardt algorithm, so to achieve a sufficient decrease, the Wolfe curvature condition is applied. Inside each Levenbeg–Marquardt step, an internal iteration is performed to obtain the descent inequality, so local minimization is required. At this stage, `nlr` computes the cubic interpolation, but if there is a computation problem, the parabolic interpolation is calculated. For more details see (Nocedal and Wright 2006, Ch. 3).

7.4 Convergence Criteria

During the iterations, if the change in the θ values is small enough, it is assumed the convergence has been achieved and iterations stop. But more precise computation is possible by using the mathematical development criterion proposed by Dennis Jr and Welsch (1978). When we have a set of data, the objective function is the sum of the function model. For $w(\theta) = \sum_{i=1}^{n} r_i^2$, the parameter θ is p-dimensional and the solution equation is

$$
\nabla w(\theta)|_{p \times 1} = \frac{\partial w}{\partial \theta}
$$

$$
= \sum_{i=1}^{n} r_i \frac{\partial r_i(\theta)}{\partial \theta}
$$

$$
= J' r
$$

$$
= 0 \; (p \text{ equation})
$$

where r is a vector of r_i values and J is the gradient matrix of derivatives of r with respect to θ. During the iteration process, the parameter θ must move to the minimum of the objective function curve in p-dimensional space, which tends to the solution of a system of linear equations. This happens when the vector r is perpendicular to the column space spanned by J, $r \perp C(\nabla w)$, so the vector r must be perpendicular to the projection of r into the column space spanned by J. This means that the cosine of degree between the projection and vector r must be small; that is (Seber and Wild (2003)):

$$
r_p = P_{\nabla w}(r) = \nabla w (\nabla w^T \nabla w)^{-1} \nabla w^T r
$$

$$
Cos(\alpha) = \frac{r^T r_p}{||r|| \cdot ||r_p||} < \epsilon \tag{7.4}
$$

where r_p is the projection of r into the column space spanned by J.

If the iteration steps are very small, the convergence condition (7.4) might not be achieved. In this case, relative precision will be used. The iteration can be stopped if either condition (7.4) holds or if the relative precision is small enough; that is:

$$
\frac{||w(\theta^{(a+1)}) - w(\theta^{(a)})||}{w(\theta^{(a+1)}) w(\theta^{(a)})} < \epsilon
$$

7.5 Mixed Algorithm

Next we combine the preceding three methods to obtain a more precise algorithm.

Algorithm 7.1 Mixed algorithm.

Step 1: Compute the model function w, gradient ∇, and Hessian H.

Step 2: Check if the Hessian is nonsingular.

    a) Nonsingular case

        2.a1: Use the Newton increment and compute $\theta^{(a+1)}$, $w^{(a+1)}$

        2.a2: If $w^{(a+1)} < w^{(a)}$ that is w is decreased

        2.a3: Otherwise iterate by the steepest descent until w decreases.

    b) Singular case

        2.b1: Use Levenberg–Marquardt iteration until w decreases.

Step 3: Check the convergence criterion. If convergence is met then the process ends, otherwise repeat step 1 with $\theta^{(a+1)}$.

7.6 Robust M-estimator

In order to find the M-estimate in the equation, the objective function must be minimized:

$$w(\theta) = \sum_{i=1}^{n} \rho\left(\frac{r_i(\theta)}{\sigma}\right)$$

The gradient and Hessian can be written as

$$\nabla = \frac{1}{\sigma} J^T(\theta) v(\theta)$$

and $H(\theta) = \dfrac{1}{\sigma^2}[J^T(\theta) D(\theta) J(\theta)] + A(\theta)$

respectively, where

$$J(\theta) = \frac{\partial r}{\partial \theta}$$

$$\frac{\partial r}{\partial \theta} = -\nabla f(\theta)$$

$$v(\theta) = \left[\dot{\rho}\left\{\frac{r_1(\theta)}{\sigma}\right\}, \dots, \dot{\rho}\left\{\frac{r_n(\theta)}{\sigma}\right\}\right]^T$$

$$D(\theta) = diag\left[\ddot{\rho}\left\{\frac{r_1(\theta)}{\sigma}\right\}, \dots, \ddot{\rho}\left\{\frac{r_n(\theta)}{\sigma}\right\}\right]^T$$

$$A(\theta) = \frac{1}{\sigma} \sum_{i=1}^{n} \dot{\rho} \left\{ \frac{r_i(\theta)}{\sigma} \right\}^T \frac{\partial^2 r_i(\theta)}{\partial\theta\partial\theta^T}$$

$$\frac{\partial^2 r_i(\theta)}{\partial\theta\partial\theta^T} = -\frac{\partial^2 f_i(\theta)}{\partial\theta\partial\theta^T}$$

$\dot{\rho}$ and $\ddot{\rho}$ are the first and second derivatives of ρ. A direct analogue for the stopping rule is:

$$\frac{v^T v_p}{||v||.||v_p||} < \epsilon$$

where $v_p = J(J^T J)^{-1} J v$.

7.7   The Generalized M-estimator

To find the generalized M-estimate in the equation, the objective function must be minimized:

$$w(\theta) = \sum_{i=1}^{n} \rho \left(\frac{Ry_i - Rf(x_i; \theta)}{\sigma} \right)$$

Since the generalized M-estimate is solved by transforming the regression model, the only value that changes is the residual $r_i(\theta) = Ry_i - Rf(x_i; \theta)$. The gradient and the Hessian are

$$\frac{\partial r}{\partial \theta} = -R \nabla f(\theta)$$

and $\dfrac{\partial^2 r_i(\theta)}{\partial\theta\partial\theta^T} = -R\dfrac{\partial^2 f_i(\theta)}{\partial\theta\partial\theta^T}$

respectively. The rest of the equation values are the same as in Section 7.6.

7.8   Some Mathematical Notation

Suppose the elements of a matrix B can be represented by b_{ij} in an appropriate dimension:

$$B_{m \times n} = [b_{ij}] \ i = 1, \dots, m \ j = 1, \dots, n$$

The three-dimensional array W can be written as

$$W_{m \times n \times p} = [w_{ijk}] \ i = 1, \dots, m, \ j = 1, \dots, n \ k = 1, \dots, p$$

The Hadamard product, denoted by (), is the elementwise product of two matrices; that is,

$$A_{m \times n} \circ B_{m \times n} = [a_{ij} b_{ij}] \ i = 1, \dots, m \ j = 1, \dots, n$$

The dotted notation is used to split a vector or matrix in the dimension where the dot is placed; that is,

$$A_{\bullet j} = \begin{bmatrix} a_{1j} \\ \vdots \\ a_{mj} \end{bmatrix}$$

and $A_{i\bullet} = [a_{i1}, \ldots, a_{in}]$.

The three-dimensional product of an array in a vector, defined by Bates and Watts (1980), is denoted by [ ] $\otimes$ [ ], and is defined as:

$$[V]_{n \times 1} \underset{n}{\otimes} [w]_{p \times p \times n} = [a_{ij}]_{p \times p},$$

$$a_{ij} = \sum_{k=1}^{n} v_k w_{ijk}.$$

7.9 Genetic Algorithm

The genetic algorithm is a method of optimization that searches in an area of predictor space. It is an efficient fast algorithm in which we try to resolve unknown initial-value situations. As we observed in nonlinear regression and its robust methods, the computation over the search area is not a well-behaved process. Far initial values might create a search area that is not in the range of a badly behaved nonlinear model. To deal with this, we provide tools in the `nlr` package that enable the user to control the search area.

The genetic algorithm in the `nlr` package is used to compute the least trimmed square (LTS) estimator (Section 3.7). In Figure 3.3, although the result from the genetic algorithm and classical optimization of the Nelder–Mead method look different, in this simulation example the percentage of outliers is high. With fewer outliers the results of the two optimization methods are the same, but for a high percentage of outliers the genetic algorithm is better. Note that LTS is not efficient and not unique, and therefore is not the final target estimate, but is used to give the initial value for the MM-estimate to achieve high breakdown and a highly efficient estimate.

8

nlr Package

8.1   Overview

A new R-package, called `nlr`, has been used to implement the methods detailed in in this book. It can be download free from the Comprehensive R Archive Network (CRAN). `nlr` is an object-oriented package system that fits nonlinear regression models using robust methods. It includes features to handle heteroscedasticity of errors, autocorrelated errors, and outlier detection. The object system can define a nonlinear regression model, a robust loss rho function, and a heteroscedastic variance in an object called `nl.form`.

`nlr` also includes features that can compute a parameter covariance matrix, plot a fitted model, predict responses, compute prediction intervals, and perform further inferences. The final fit results are accessible through output objects that are flexible, so that researchers can apply them to their own particular problems.

To use the `nlr` package, we first have to define a nonlinear regression function model $y = f(\mathbf{x}_i; \theta)$ as an `nl.form` object. This is an extension of the initial object definition by Bunke et al. (1995b), with extra functionality. The data set for which the model is going to be fitted has to be adjusted to use the names of the response and predictor variables defined in a nonlinear function model object. A robust ρ function should also be defined as an `nl.form` object. Furthermore, a heterogeneous variance function, $H(x_i; \tau)$ or $\sigma^2 g(x_i; \lambda)$, can be defined as an `nl.form`. This can cover all three cases in Chapter 4, Equations (4.2)–(4.5).

The `nlr` function fits the model using methods that depend on the choice of classical or robust options. Depending on the fitting method used, the result is stored in different kinds of objects, in an inheritance scheme. This includes all the information used in the procedures. The final fitted object can be used for inference purposes through several methods implemented in the `nlr` package for a specified output object result. A brief example of the procedure follows.

Robust Nonlinear Regression: with Applications using R, First Edition.
Hossein Riazoshams, Habshah Midi, and Gebrenegus Ghilagaber.
© 2019 John Wiley & Sons Ltd. Published 2019 by John Wiley & Sons Ltd.
Companion website: www.wiley.com/go/riazoshams/robustnonlinearregression

Example 8.1 *Cow milk data*
Consider the cow milk data discussed in Chapter 2 and shown in Table A.8. The data follow the Wood model (2.22), $f(x; \theta) = ax^b e^{-cx}$. First, we define this model function as an nl.form object. The function model can be defined as an R expression, in which the gradient and Hessian matrices with respect to the parameter vector $\theta = (a, b, c)$ can be derived by the derive3 function, called convexpr2nlform.

```
>wood = convexpr2nlform(yr ˜ a * xr ^ b
        * exp(-c * xr),
        start = list(a = .05, b = 4.39,
        c = 21.6))
```

To fit the model using the ordinary least squares method the nlr function is

```
>fittcow <- nlr(
  wood,
  data = list(xr = nlr::cow$Day,
  yr = nlr::cow$Milk),
  start = list(a = 11, b = .3, c = .003),
  control = nlr.control(method = "OLS")
)
```

The plot command is shown in Figure 2.1.

```
> plot(fittcow,
       control=nlr.control(length.out=30))
```

The parameter estimate values are saved in the parameters variable and the following command displays the parameter estimate values, which are precisely a = 11.26233, b = 0.3378616, and c = 0.00218526):

```
> fittcow$parameters
```

The computed scale value ($\sigma = 1.674667$) can be accessed by

```
> fittcow$scale
```

8.2 nl.form Object

nl.form is a general object that can store a nonlinear regression function model $f(\mathbf{x}_i; \theta)$, a robust rho function model $\rho(t)$, or a nonlinear heterogeneous function $G(x; \tau)$. It can store an R-expression or an R-function. Moreover, it can include nonlinear parameters of function models, like θ, or heterogeneous variance function parameters τ, independent variables **X**, dependent variables **Y**, tuning constants of robust rho ρ functions, and any other parameters. Since nlr is a gradient-based package, the gradient and Hessian with respect to parameters of both the right- and left-hand sides of the regression model should be stored as attributes. The nl.form object can be created by an initializer as:

```
new (
    "nl.form",
    formula,
    fnc,
    formtype,
    p,
    inv = NULL,
    name = name,
    par,
    arguments = list(...),
    dependent,
    independent,
    origin,
    selfStart
)
```

An easier way is to use a contractor function in which slot types can be identified automatically; that is,

```
nl.form (
    form,
    p = NULL,
    inv = NULL,
    name,
    par = NULL,
    dependent = NULL,
    independent = NULL,
    origin = NULL,
    selfStart = NULL,
    ...
)
```

Here the arguments are defined as follows:

form: Stores mathematical model function. Can be defined as a `call`, `expression`, `formula` or `function` R object.

p: Number of parameters.

inv: Inverse of the nonlinear function model, designed for feature development, but currently is not functional and in most objects is omitted.

name: Character name for the function model.

par: `list` objects of parameter values, including variables with the same name as the parameter.

dependent: Single character or list vector of character names of dependent variables used in the model.

independent: Single character or list vector of character names of independent variables used in the model.

selfStart: Self-start function for computing the initial values of parameters.

`origin`: Original form of function expression. This is the original expression that might be sent to the `convexpr2nlform` function to calculate the gradient and Hessian.

... Any extra argument passed to the `nl.form` internal function or `nlr` function.

An R `function` is stored in the `fnc` slot of the `nl.form` object, while the `formtype` slot is for the "function" character. Other types of model object can be stored in `formula`, and the `formtype` slot must be equal to "formula" for "expression", "call" or "formula" R types. The model can be a two-sided expression formula defined as the "response model", including the predictor function and response, or a one-sided formula expression defined as the "model". The Hessian and gradient can be included as attributes on each side. For simplicity, the function `convexpr2nlform` is implemented to derive expressions for the constructed Hessian and gradient. It calls the `derive3` function from the MASS library, so the functions that are defined in `derive3` can be used inside a model. Meanwhile, for undefined functions in `derive3`, a direct expression or function can be defined by the user manually.

Example 8.2 *Define an R `Expression` as an `nl.form` object*
In Example 8.1 the Wood model is defined as:

```
>wood = convexpr2nlform(yr ~ a * xr^b * exp(-c * xr),
              start = list(a =.05, b = 4.39, c = 21.6),
              name = "Wood")
```

and the `nl.form` object created includes the following slots:

```
> wood$formula
yr ~ expression({
    .expr3 <- exp(-b * xr)
    .expr6 <- exp(-c * xr)
    .expr7 <-.expr3 -.expr6
    .expr9 <-.expr3 * xr
    .expr11 <-.expr6 * xr
    .value <- (a) *.expr7
    .grad <- array(0,
            c(length(.value), 3L),
            list(NULL,
                c("a","b", "c")))
    .hessian <- array(0,
                c(length(.value), 3L, 3L),
                list(NULL,c("a", "b", "c"),
                    c("a", "b", "c")))
    .grad[, "a"] <-.expr7
    .hessian[, "a", "a"] <- 0
    .hessian[, "a", "b"] <-.hessian[, "b", "a"]
                        <- -.expr9
    .hessian[, "a", "c"] <-.hessian[, "c", "a"]
```

```
                      <-.expr11
.grad[, "b"] <- -((a) *.expr9)
.hessian[, "b", "b"] <- (a) * (.expr9 * xr)
.hessian[, "b", "c"] <-.hessian[, "c", "b"]
                      <- 0
.grad[, "c"] <- (a) *.expr11
.hessian[, "c", "c"] <- -((a) * (.expr11 * xr))
attr(.value, "gradient") <-.grad
attr(.value, "hessian") <-.hessian
.value
})
```

The general form of the `convexpr2nlform` function is

```
convexpr2nlform(form, start,name)
```

where the `form` argument is a one- or two-sided R expression that includes parameters, and dependent and independent variables.

The `start` argument is a list of parameters, which are the initial guess for the parameter values that will be stored in the `par` slot of the derived `nl.form` object. Although the `par` slot stores some values for parameters, globally they are not good initial values for any example. In `nlr`, there are several options to define initial values, but in the rare situation where initial values cannot be defined in any of these ways, the only choice will be the values stored in the `par` slot. It is compulsory to define parameter values and the names for each of the parameters in this slot.

The `name` argument is the character of the name of the model. It is stored in the `name` slot of the `nl.form` object. The constructed object returns a two-sided expression, in which the left-hand side calculates the response value or a general function of independent variable yr. The right-hand side calculates the predictor function Wood model and its gradient and Hessian with respect to the parameter vector $\theta = (a, b, c)$, which is attached as attributes of the computed values. The gradient of the Wood function is defined as

$$\left.\frac{\partial f(x;\theta)}{\partial\theta}\right|_{\theta=(a,b,c)} = \begin{bmatrix} x^b \exp(-cx) \\ ax^b \ln(x) \exp(-cx) \\ -ax^{b+1} \exp(-cx) \end{bmatrix}$$

with the Hessian matrix equal to

$$\left.\frac{\partial^2 f(x;\theta)}{\partial\theta\partial\theta^T}\right|_{\theta=(a,b,c)} = \begin{bmatrix} 0 & x^b \ln(x)e^{-cx} & -x^{b+1}e^{-cx} \\ x^b \ln(x)e^{-cx} & ax^b(\ln(x))^2 e^{-cx} & -ax^{b+1}\ln(x)e^{-cx} \\ -x^{b+1}e^{-cx} & -ax^{b+1}\ln(x)e^{-cx} & ax^{b+2}e^{-cx} \end{bmatrix}$$

The computed value of the `wood` object is a list that includes two vectors on the right- and left-hand sides of the expression. The left-hand result is called the `response` and is a vector of response yr, while the right-hand side is called the predictor and is a vector of computed function *f*, including a gradient attribute matrix of gradient values, of size $n \times p$, and a hessian attribute array array of Hessian values, of size $n \times n \times p$. The object can be computed using the eval method of the nl.form object for the values of the cow milk data and at the starting values as:

```
>cowvalues<-eval(wood,list(xr=nlr::cow$Day,yr=nlr::cow$Milk,
               a=11,b=.3,c=.003))
```

We can access the computed response for the left-hand side by:

```
>cowvalues$response
```

We can access the computed predictor values for the right-hand side by:

```
>as.numeric(cowvalues$predictor)
```

The 10×3 gradient matrix value is accessed by

```
>attr(cowvalues$predictor,"gradient")
```

and the $10 \times 10 \times 3$ array of Hessian values is accessed by

```
>attr(cowvalues$predictor,"hessian")
```

The nl.form object can include a user-defined function. For this purpose the form argument of the nl.form constructor can be a function. For computational purposes, the user-defined function should return the gradient and Hessian derivatives with respect to parameters as attributes in "response" or "predictor" variables. It is possible to define any extra arguments, such as tuning constants, of the user-defined function and pass them to the nlr procedures.

Example 8.3   Suppose we want to define the Hampel function (denoted by $\rho(t)$), shown in Table A.11, as an nl.form object. It will be a one-sided function with dependent variable (t) and a few tuning constants as argument entries. The function has to return first ($\psi = \rho'$) and second ($\psi' = \rho''$) derivatives with respect to t, and the weight function is defined by (1.25). The constructor is:

```
hampel <- nl.form(
  form = function(t,
                  a = 1.345,
                  k0 = 3.73677,
                  k1 = 4,
                  maxrho5 = 1.345,
                  ...) {
    U <- abs(t)
    Ugrta <- (U > abs(a))
    .rho <-.grad <-.hess <-.weight <- NULL
    .rho[Ugrta] <- 2. * abs(a) * U[Ugrta] - a * a
    .rho[!Ugrta] <- t[!Ugrta] ^ 2
    .grad[Ugrta] <- 2. * abs(a) * sign(t[Ugrta])
    .grad[!Ugrta] <- 2. * t[!Ugrta]
```

```
    .hess[Ugrta] <- 0.
    .hess[!Ugrta] <- 2.
    .weight[Ugrta] <- 2. * abs(a) / U[Ugrta]
    .weight[!Ugrta] <- 2.
    attr(.rho, "gradient") <-.grad
    attr(.rho, "hessian") <-.hess
    attr(.rho, "weight") <-.weight
    return(.rho)
  },
  name = "huber",
  independent = "t",
  a = 1.345,
  k0 = 3.73677,
  k1 = 4,
  maxrho5 = 1.345
)
```

where `t` is the function variable, and `a`, `k0`, `k1`, `maxrho5` are tuning constants that are defined with default values. These tuning constants can be sent as an "…" argument to several functions in `nlr`. The derivatives $\psi = \rho', \psi' = \rho''$ and the weight $w(t) = \psi(t)/t$ are saved as attributes in the result.

Note that the `robustbase` R-package includes functions, such as `psi.hampel`, that allow users (through the `psiFunc` function) to create their own psi functions. The gradient and hessian in `robustbase::psiFunc` represent the rho, psi, and wgt arguments. The facilities in `nlr` were developed independently from the `robustbase` package. The object format `nl.robfuncs` can define a robust function and its gradient and hessian as attributes, but does not depend on `robustbase::psiFunc` of the `robustbase` package. In fact the `nlr` package started as an SPLUS package, then moved to R from 2008–2017. Facilities that mirror those in `robustbase` were therefore created unintentionally.

8.2.1 `selfStart` Initial Values

Initial values for parameters can be calculated from a geometric interpretation of the model and then inserted in the `selfStart` function slot. These values are used in optimization iterations and estimating procedures to find the final values of a parameter. When the argument `start` is omitted in the `nlr` function call, the defined `selfStart` slot function will be used.

The `nlr` package tries several ways to find initial values, but in general it is not easy to find good initial values. High initial values might cause the model and its derivatives to not non-computable, and in this case an error such as "Cannot compute the model, misdefined model or bad choice for starting point" might be shown.

The `plotinitial` function can draw two-dimensional data and the curve of the model at the initial values. Its general form is as follows:

```
plotinitial(form,
            data,
            start = getInitial(form, data),
            control = nlr.control(
                          length.out = 100))
```

The arguments are defined as follows:

`form`: `nl.form` object of the nonlinear model.

`data`: Data points to be plotted, including response and predictor variables.

`start`: Initial values of the parameters. Can be any value. If omitted, the `selfStart` function will be called.

`length.out`: Length of the predictor variables, which will be incremented to calculate their response values. Used to control the smoothness of the predicted curve. If omitted, the prediction curve will be calculated using the sample data set values, and if the number of samples is small the curve looks like a broken line and so the `length.out` argument can be used to increase the number of predicted points and draw a smoother curve.

Example 8.4 The scaled exponential convex equation (3.22b), $y_i = p_1 + e^{-(p_2 - p_3 x_i)}$, is used to model the carbon dioxide data in Section 3.14. The limit of the model at minus infinity is p_1, so the parameter p_1 can be approximated by the minimum of the data. By taking the logarithm from the model we have $\log(y - p_1) = -p_2 + p_3 x$, so the initial values for parameters p_2 and p_3 can be found by fitting the linear regression of $\log(y - p_1)$ against $-p_2 + p_3 x$.

Using this idea, the scaled exponential convex model can be defined as the `nl.form` object with `selfStart` as:

```
ScalExp <- convexpr2nlform(
    yr ~ p1 +
    exp(-(p2 - p3 * xr)),
      selfStart = function(data) {
        y1 <- as.double(data$yr)
        p1 <- min(y1) # Assume p3 be positive
        y <- log(y1 - p1 + 10 *
              .Machine$double.eps)
        x <- as.double(data$xr)
        b1 <- lm(y ~ x)
        p2 <- -b1$coefficients[1]
        p3 <- b1$coefficients[2]
        return(list(p1 = p1, p2 = p2, p3 = p3))
      },
      name = "Scaled Exp2 convex",
      start = list(p1 = 700, p2 = 21, p3 = 0.01)
  )
```

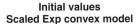

**Initial values
Scaled Exp convex model**

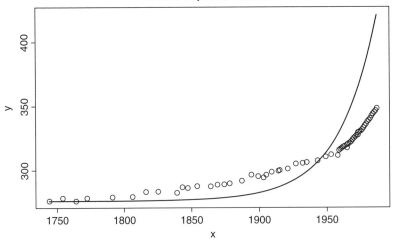

Figure 8.1 Plot of predicted values for t he scaled exponential model, computed using initial values from the selfStart slot for carbon dioxide data.

To plot the computed initial values using the selfStart function, one can use the plotInitial function (see Figure 8.1 for the carbon dioxide data example). As can be seen, the initial values are not the best answer, but they are close enough to estimate the parameter using nlr procedures. The optimum estimates will be discussed in the next section.

```
crbdt <- list(xr = nlr::carbon$year,
  yr = nlr::carbon$co2)
plotinitial(form = ScalExp, data = crbdt)
```

The ordinary least squares estimates can be computed using the following command:

```
carbon.ols <- nlr(formula = ScalExp,
        data = crbdt,
        control = nlr.control(method = "OLS"))
```

in which the defined selfStart function will be used to compute the initial values.

8.3   Model Fit by nlr

The nlr function is implemented to allow nonlinear model to be fitted using all of the methods discussed in this book. In general, nlr requires three kinds of nl.form object: a nonlinear function model $f(x; \theta)$, a heterogeneous function

$G(x_i; \tau)$, and robust loss function $\rho(.)$. The general structure of nlr function arguments is as follows:

```
nlr(
   formula,
   data = parent.frame(),
   start = getInitial(formula, data),
   control = nlr.control(),
   weights = NULL,
   robustobj = NULL,
   robustform = c(
     "hampel",
     "huber",
     "bisquare",
     "andrew",
     "halph huber",
     "hampel 2",
     "least square"
   ),
   varianceform = NULL,
   tau = NULL,
   correlation = NULL,
   covariance = NULL,
   ...
)
```

The arguments are defined as follows:

formula: Can be a nl.form object of the nonlinear function model or a model formula defined as R-Expression, with the response on the left-hand side of a ~ operator and an expression involving parameters and covariates on the right.

data: an optional data frame or list of data with the response and predictor as the names of variables. If the heterogeneous case includes response variable values of the heterogeneous variance function, nlr assumes the variance model is an explicit function of predictor x_i; that is, $G(x_i; \tau)$. Otherwise it assumes that it is a function of predictor $G(f(x_i, \theta), \tau)$. In this case, the predicted values $f(x_i, \hat{\theta})$ will be computed and the variance model will be calculated at the predicted values.

start: data.frame or list object of the starting values of parameters. The name of the parameters must be represented as the names of variables in the list.

control: nlr.control list, which controls options in nlr procedures (see nlr.control).

weights: User option matrix of the variance covariance matrix of error, a general weight that can be used by the user. If a correlation, covariance or variance form argument is given, it will be ignored.

`robustobj:` =NULL, optional `nl.form` object of the robust loss function defined by the user. The user can define their own function to be an `nl.form` object or use the predefined `robloss` function in `robustform`.

`robustform:` The name of the `nl.form` object of the robust loss ρ function to be used for computing the MM-estimates. It recalls one of the defined ρ functions from the `nl.robfuncs` variable list. It can be one of a list of string names, such as "hampel", "huber", "bisquare", "andrew", "halph huber", "hampel 2", or "least square".

`varianceform:` =NULL, optional `nl.form` object of heterogeneous variance form.

`tau:` =NULL, optional list or `data.frame` of initial values for the heterogeneous variance function parameter.

`correlation:` =NULL, optional autocorrelated error form. A list of autocorrelation commands that provides capabilities to use `corStruct`, as defined in the `nlme` package, or to define the user autocorrelation defined function.

`covariance:` =NULL, optional covariance matrix of errors. If given, generalized estimates will be calculated.

`...` Any extra arguments to any function sources, such as nonlinear regression model, heteroscedastic variance function, robust loss function or optimization object function.

By default, the `nlr` function estimates parameters using robust methods based on the MM-estimator, as discussed in Section 3.11. However, the `control` argument, called the `nlr.control` function, allows the user to select estimation methods and set some characteristics of the computational algorithms. For example, `nlr.control` includes a string argument called `method` that sets the method of estimation. If it is omitted, the default value MM-estimate will be calculated. `nlr.control(method="OLS")` means that the parameters will be estimated using ordinary least squares and `nlr.control(method="lms")` means the parameters will be estimated using least median squared residual estimates. Further details of the control argument are discussed later in this chapter.

Example 8.5 Consider the scaled exponential convex in Example 8.4. To fit this model to the carbon dioxide data in Table A.3 using the ordinary least squares method, we can use the `nlr` function by setting the `formula` argument equal to the defined `ScalExp` object, and the `data` argument to a list of predictor and response values that must be named the same as the `predictor` and `response` variables in the predefined `ScalExp` object (xr, yr). We also set the `method` via setting the argument equal to "OLS" using `nlr.control(method="OLS")`.

```
crbdt<-list(xr=nlr::carbon$year,yr=nlr::carbon$co2)
carbon.ols <- nlr(formula=ScalExp, data=crbdt,
                  control=nlr.control(method="OLS"))
```

Recall that the omitted `start` argument forced the `nlr` package to calculate the initial values using the predefined `selfStart` slot of the `ScalExp` object. To access the parameter estimates, we can use `carbon.ols$parameters`. To fit the model using a robust method we can call the function by omitting the `control` argument, which by default calculates the MM-estimate and the Hampel rho function defined in the `nl.form` object in Example 8.3.

```
carbon.mm <- nlr(formula=ScalExp, data=crbdt)
```

To explicitly specify the robust loss rho function we can use the `robustform` argument to call the predefined ρ functions in `nlr` or use the `robustobj` argument to call a user-defined ρ function. For example, the following code explicitly calls the Hampel function, giving the same result as above:

```
carbon.mm  <- nlr(formula=nlrobj5[[8]], data=crbdt,
         robustform ="hampel")
```

The `robustform` argument calls the predefined loss functions from the `nl.robfuncs` variable list in `nlr` or the user-defined `nl.form` object robust function. The character name of the list argument that specifies the function can be one of these values: "hampel","huber", "bisquare","andrew", "halph huber", "hampel 2", "least square" (see Table A.11).

8.3.1 Output Objects, `nl.fitt`

The result from the `nlr` function is an object that depends on the estimation method called (see Figure 8.2). The `nl.fitt` object stores the "OLS" fit and the `nl.fitt.gn` object, inherited from `nl.fitt`, stores the generalized

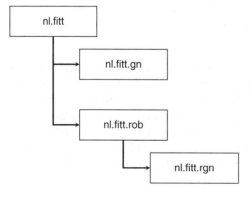

Figure 8.2 Output objects, inheritance hierarchy.

least squares estimator, including heteroscedastic or autocorrelated error cases. The `nl.fitt.rob` object, also inherited from `nl.fitt`, stores the estimates from robust methods, specifically the MM-estimate. `nl.fitt.rgn`, inherited from `nl.fitt.rob`, stores the generalized robust estimate in heteroscedastic or autocorrelated error cases. The `nl.fitt` object slots hold the data parts and methods. Data slots store the numeric values of the result. These values can be accessed by the $ sign. Methods are functions that perform a physical job. For example, `plot` can display the data points and the fitted curve. The `nl.fitt` data (D) and methods (M) are as follows:

Slots of `nl.fitt` object:

`parameters:`   (D) List of estimated parameter values $\hat{\theta}$, which might contain the scale value σ but not for all the results from `nlr`. To access the scale, the `scale` slot can be used.

`scale:`   (D) Computed scale $\hat{\sigma}$, the constant standard error of the model error ε.

`correlation:`   (D) Correlation of the fitted model.

`form:`   (D) `nl.form` object of the nonlinear function model, which is originally called by `nlr`.

`response:`   (D) Computed values of left-hand side response values, typically the response variable values y_i, but in some applications may be a function of response variables.

`predictor:`   (D) Computed values of right-hand side predicted values, typically $f(x_i; \hat{\theta})$. The gradient and Hessian are computed and attached as attributes, so we can access an $n \times p$ matrix of the gradient by `attr(ObjectName $predictor,"gradient")` and an $n \times n \times p$ array of the Hessian by `attr(ObjectName$predictor, "hessian")`.

`curvature:`   (D) List of parameter-effect (pe) curvature and relative intrinsic curvatures, as defined by Bates and Watts (1980). Indicates the adequacy of a linear approximation and its effect on inference.

`history:`   (D) Matrix of the history of iteration procedure values. The rows of the matrix are iterations, and columns include parameter values, sums of squares, and convergence criterion.

`method:`   (D) Object of type `fittmethod`, defined in `nlr`, representing the methods and procedures used in the computation procedure.

`data:`   (D) List of the sample data called by the user and used in procedures.

`sourcefnc:`   (D) Source function called by the user. Might not be called directly by the user.

`Fault:`   (D) Object of type `Fault`, predefined in `nlr`. If any error happens during the computation procedure it will be stored in the `Fault` object. It contains error codes (stored in the `nlr::db.Fault` table variable) and the function name of the error that occurred.

others: (D) Other values that might be computed in procedures.

$: (M) Used to access the slot values.

residuals(nl.fitt,data=NULL): (M) Calculates residuals, $r_i = y_i - f(x_i, \hat{\theta})$, $i = 1, \dots n$. The data can be a list of data values with the same name as the predictor values, given by the user. The residuals will be calculated at the given values. If this argument is omitted, the residuals of the original data set will be calculated.

predict(nl.fitt, data=NULL): (M) Calculates the prediction values $f(x_i, \hat{\theta})$. The data can be a list of data values with the same name as the predictor values, given by the user. The prediction will be calculated at the given values. If this argument is omitted, the prediction of the original data set will be calculated.

hat(nl.fitt): (M) Computes the hat matrix. For extending the hat matrix to nonlinear regression, the gradient of the nonlinear function model is used as the design matrix in the linear regression model hat matrix formula.

plot(nl.fitt, control=nlr.control(history=F,length.out=NULL,singlePlot=F)): (M) Plots the data and curve fit in two-dimensional regression, plus the residual plot. The history=T argument causes the plot function to draw the convergence of the procedure. The length of the predictor variables will be incremented to calculate the response values. The length.out argument is used to control the smoothness of the predicted curve. If this argument is omitted, the prediction curve will be calculated at the sample data set values, and if the number of samples is small the curve will look like a broken line. The length.out argument can be used to increase the number of predicted points and draw a smoother curve (see Example 8.6).

atypicals(nl.fitt): (M) Used to identify atypical points. Calculates the statistical measures for identifying the outliers, for example studentized residuals, the Cook distance, the Mahalanobis distance, the Hadi potential, and minimum volume ellipsoid (MVE) estimate. These are calculated based on the gradient of the predictor values and the Jacobian leverage, which has the prefix "jl".

atypicals.deleted(nl.fitt): (M) Used to identify atypical points. Calculates the statistical measures for identifying outliers based on deleting a point at a time, for example $\hat{y}_i$, deleted studentized residuals, DFFITS, the Atkinson distance, and DF-Betas. They are calculated based on the gradient of the predictor values and Jacobian leverage, which has the prefix "jl".

recalc(nl.fitt,data=NULL): (M) Recalculates the model fit at the new point data. Implemented for feature development.

JacobianLeverage(nl.fitt): (M) Calculates Jacobian leverage.

parInfer(nl.fitt, confidence=0.95): (M) Calculates measures for estimated parameters for inference purposes, such as the covariance matrix, correlation matrix, standard errors, and confidence intervals, with probability equal to the given argument `confidence`.

`predictionI(nl.fitt,confidence=.95, data=NULL):` (M) Calculates the prediction interval with probability equal to the given argument `confidence`. If the data are omitted the prediction interval at the original data point will be computed.

`acf(nl.fitt, lag.max = NULL, type = c("correlation", "covariance", "partial"), plot = TRUE, na.action = na.fail, demean = TRUE, ...):` (M) Calculates the autocorrelation and appropriate plots for the computed residuals. The other arguments are the same as the `stats::acf` function in R.

(Note: `nl.fitt` is the name of the output result variable assigned by the user.)

Example 8.6 The fitted scale exponential convex model for the carbon dioxide data in Example 8.5, stored in the `carbon.ols` variable, is in the `nl.form` object. Thus, all the slots and methods given above are accessible. For example, the command

```
plot(carbon.ols,control=nlr.control(history=T))
```

draws the fitted curve, the residuals and the history of convergence (see Figure 8.3). The residual plot shows the data are autocorrelated, or that the fitted model might not be suitable. Attempts to identify the correlation structure were unsuccessful, so the problem is still open.

8.3.2 Output Objects, `nl.fitt.gn`

The `nl.fitt.gn` object is used to save the results from the model when there is heteroscedastic variance or autocorrelated errors. These cases can be fitted by generalized least squares estimates. The `nl.fitt.gn` object is inherited from `nl.fitt` and therefore includes all the slots given in the previous section plus some others.

Slots for `nl.fitt.gn` object:

`vm:` (D) Generalized covariance matrix V, where $E(\varepsilon) = \sigma^2 V$. Covers the covariances from both the heteroscedastic and autocorrelated error cases.

`vm:` (D) The R matrix is the Cholesky decomposition of $V = R^T R$.

`hetro:` (D) The results from the "CME" and "CLsME" methods for modeling the heteroscedastic variance case will be saved in the `hetro` slot. The heteroscedastic variance model, $G(x_i; \tau)$, can include unknown parameters τ in general cases (see Chapter 4).

`autcorr:` (D) The results from the "TS" method for modeling the autocorrelated errors case will be saved in the `autcorr` slot.

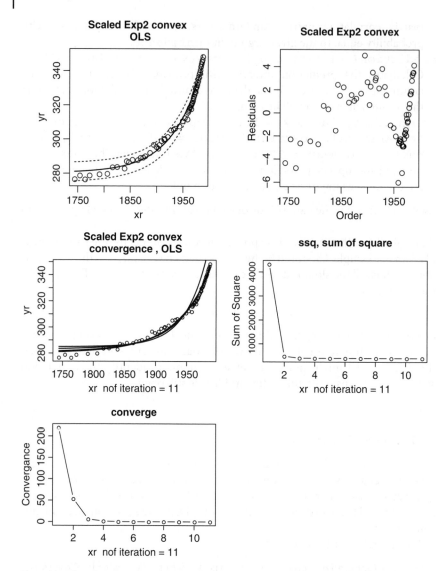

Figure 8.3 Fitted scale exponential convex model for the carbon data, using the `history=T` command.

`autpar:` (D) Autocorrelation structure parameter fits.

`gresponse:` (D) Transformed response variable $R\mathbf{Y}$, including the transformed gradient and Hessian attributes.

`gresponse:` (D) Transformed prediction values $R\eta(\hat{\theta})$, including the transformed gradient and Hessian attributes.

8.3.3 Output Objects, `nl.fitt.rob`

By default, `nlr` uses the MM-estimate method, discussed in Section 3.11, for estimating parameters for homogeneous variance and independent error cases. The result is saved in the `nl.fitt.rob` object, which is inherited from `nl.fitt`, so it includes all slots and methods from `nl.fitt` plus extra slots for saving the robust fit objects.

Slots of `nl.fitt.rob` object:

`htheta`: (D) Single value of the minimized robust loss function

$$h(\theta) = \sum \rho \left(\frac{r_i(\hat{\theta}_{MM})}{k_1 \hat{\sigma}_n} \right)$$

The $p \times 1$ vector of the gradient $\dot{h}(\hat{\theta})$ and the $p \times p$ matrix of the Hessian $\ddot{h}(\hat{\theta})$ are saved as "gradient" and "hessian" attributes, respectively, in `htheta`.

`rho`: (D) Vector of robust loss values of $\rho(r_i(\hat{\theta}_{MM})/(k_1 \hat{\sigma}_n))$, including the first derivatives ψ function, second derivatives ψ' and weights as "gradient", "hessian" and "weight" attributes, respectively.

`ri`: (D) Vector of calculated residuals $r_i(\hat{\theta}) = y_i - f(x_i, \hat{\theta})$.

`robform`: (D) `nl.form` object of called ρ function.

`JacobianLeverage`: (M) The same slot as in `nl.fitt` but the robust Jacobian leverage will be calculated.

8.3.4 Output Objects, `nl.fitt.rgn`

The `nl.fitt.rgn` object is used to save the results from the model with heteroscedastic variance or autocorrelated errors. These cases can be fitted by the generalized MM-estimate. The `nl.fitt.gn` is inherited from `nl.fitt.rob`, and therefore includes all its slots plus some others.

Slots of `nl.fitt.rgn` object:

`vm`: (D) Generalized covariance matrix V, where $E(\varepsilon) = \sigma^2 V$. Covers the covariances from both the heteroscedastic and autocorrelated error cases.

`vm`: (D) The R matrix is the Cholesky decomposition of $V = R^T R$.

`hetro`: (D) The results from the "RME" and "RGME" methods for modeling the heteroscedastic variance case will be saved in the `hetro` slot. The heteroscedastic variance model, $G(x_i; \tau)$, may include unknown parameters τ in the general case (see Chapter 4).

`autcorr`: (D) The results from the "RTS" method for modeling the autocorrelated errors case will be saved in the `autcorr` slot.

`autpar`: (D) Autocorrelation structure parameter fits.

`gresponse`: (D) Transformed response variable RY, including the transformed gradient and Hessian attributes.

`gresponse`: (D) Transformed prediction values $R\eta(\hat{\theta})$, including the transformed gradient and Hessian attributes.

8.4 nlr.control

The nlr.control argument of the nlr function controls all the estimation, optimization methods, computation algorithms, plot arguments, and initial values. Any of the arguments may have applications in one or more functions but not all; in other words, not all arguments are necessarily used in all functions. The nlr.control object makes the computation derivative free or, as the default, derivative based. However, derivative-free methods are not efficient in nlr. An appropriate model structure, such as for the heterogeneity or autocorrelated cases, can be defined by the nlr function arguments and not the nlr.control argument values, which will be explained in the rest of this chapter. The nlr.control arguments are given below.

maxiter=50 : Maximum number of iterations. If the number of iterations exceeds this value, the output object will include a fault slot with fault number FN = 1. This does not necessarily mean the algorithms are not converging, and the user can increase the values or change initial values and test the result.

tolerance = 0.0001 : Tolerance of convergence. Each optimization algorithm has a different value. Generally the convergence criteria discussed in Chapter 7 will be compared with this value.

minlanda = 1e-16 : Minimum value of λ allowed in the steepest descent optimization algorithm.

trace = F : Traces the computation iterations by drawing the iteration curve of each iteration. At the moment, this function is only operational in the MM-estimate procedure in the two-dimensional case. Users can trace the procedure to identify how its initial value and the iterations proceed, but the design is not efficient in this version.

derivfree=F : Specifies whether the computation should be derivative-free (T) or derivative-based (F). By default, all the computations are derivative-based. The derivative-free methods, based on the Nelder–Mead optimization algorithm, are inefficient and exhibit slow convergence. If a derivative-free computation is required, the maxiter value can be increased, possibly by more than hundreds.

robscale=T : If this is set to F, the robust M-scale will be computed during the MM-estimate, otherwise the robust M-scale estimate will be computed for scale values. In different versions of the package, different values are used for testing purposes, so it is not recommended to change the default value.

algorithm = c("Levenberg–Marquardt", "Nelder-Mead", "Gauss Newton"): Only one of the values can be used. The NULL or default value is Levenberg–Marquardt. The Nelder–Mead or derivfree=F arguments have same effect, specifying use of derivative-free Nelder–Mead optimization methods. The Gauss–Newton method was used in older versions of the package, but is not functional in this version. It has been slated for further development.

initials=c("manual","lms","OLS","quantile"): Only one of the values will be used. Specify the initial values for the starting point of the parameters. The default value is "manual" which gets values from the start argument. The "lms" value calculates the least median square (LMS) estimate (Chapter 3). This estimate is not unique and might be far from the actual value, but if it gives a value it is high breakdown point and is an appropriate starting value for robust methods. "OLS" computes the ordinary least square estimate, which is not robust and not recommended for robust methods. The "quantile" value computes the quantile regression from the quantreg package. It reduces the efficiency of the computation but it can be a good choice if the user does not know the initial values. The first priority for initial values in the nlr function is the initials argument, as mentioned in nlr.control. If this argument is committed, the values the user gives to the nlr function call's start argument will be considered. If the start argument is omitted, first nlr searches in the data argument with the same name as the parameter names. If the data argument does not contains parameters, the selfStart function in the nl.form object will be called. If the selfStart function does not exist the values of the parameters in the par slot in the function model nl.form object will be used.

method=c("default", "RME", "CME", "CLSME", "RGME", "WME", "MLE", "OLS", "TS", "RTS", "lms"): Method of estimation for the nlr function. First identify which cases of the weighted, heterogeneity, autocorrelated error, and general covariance structure model should be fitted. This can be done from the given weights, variance-form, correlation, and covariance arguments (see nlr function structure). After that, the model can be fitted using the method given in the method argument given in nlr.control. The default value when the weights and covariance argument is given in the nlr function is a robust MM-estimate. For the heterogeneous case when the variance-form argument is given in the nlr function, the RME method will be used. For the heterogeneous case when the correlation argument is given in the nlr function, the RTS method will be used. Meanwhile, in each case the derivfree argument specifies if derivative-free or derivative-based approaches should be used.

history = F: Used in a plot to force plot the iteration history.

length.out = NULL Output length for the *x*-coordinate increment in the plot function. Used to create a smoother plot if the number of *x* values is low.

singlePlot = F: plot in nlr draws the fitted plot and residual plot in two columns. If singlePlot = T they will be drawn in different windows.

singularCase = 1: Chooses the singularity case selection method. The default singularCase=1 uses all eigenvalues to remedy the singularity in Levenberg-Marquardt iteration (Equation 7.3). singularCase=2 use the constant eigenvalue (Equation 7.2).

JacobianLeverage = c("default", "classic", "robust"): Applied in the atypicals function for computing outlier detection measures, "default" or "robust" values, and by default computes the classical Jacobian leverage (Equation 6.9) for OLS, and robust Jacobian leverage (Equation 6.17) for a robust fit. The "classic" value forces atypicals to perform the computation of robust estimates by classical Jacobian leverage instead of robust Jacobian leverage.

8.5   Fault Object

If any error or warning occurs during an nlr procedure, the output includes the Fault object, which explains the error type and the function that produced the error. The Fault object is mostly used for development purposes, but it might help the user to identify the null condition and to correct it, for example if there are missing data or bad initial values. The user can send this object to the authors to get help to find a solution for a problem.

Slots of the Fault object:

FL=F: (D) True value indicates that an error occurred.

FN=0:   (D) Fault number, which indicates the error code. The error codes are stored in the db.Fault variable.

FT: (D) Text of the error explains the error.

FF: (D) Name of the function that raised the fault.

is.Fault(object):   (M) Returns true if an error occurred. The "object" can be a Fault object or any of the output objects.

is.Faultwarn(object):   (M) Returns true if a warning (nonzero FN) occurs.

is.Warn(object):   (M) Returns true if a warning occurred but not an error.

8.6   Ordinary Least Squares

As shown in Example 8.5, ordinary least squares estimates can be computed by calling the control=nlr.control(method="OLS") argument in the nlr function. nlr uses the Gauss–Newton algorithm, as given by Bates and Watts (2007), when derivatives exist and the Nelder–Mead algorithm for the derivative-free method. OLS is sensitive to initial values: far-away initial values might result in unknown values for the nonlinear function model. Meanwhile the Nelder–Mead algorithm is worse, as it not only depends on initial values but also on the incremented interval over which it searches for the optimum. To fit a model using the Nelder–Mead algorithm, if we start from initial values $\theta = (\theta_1, \ldots, \theta_p)$, we need p more points to be incremented by, for example, $\delta = (\delta_1, \ldots, \delta_p)$, each point once $(\theta_1 + \delta_1, \ldots, \theta_p), \ldots, (\theta_1, \ldots, \theta_p + \delta_p)$. The increment δ can be adjusted by the `delta` argument in the … of the `nlr` function. If it is omitted, the δ will be set to $0.5 \times \theta$ by default. However, the following example shows that such a default is not appropriate, even if we use the `selfStart` initial value. The value should therefore be set by the user to a small value, so as to control the parameters to be close to the initial values. Again such an adjustment requires knowledge of the true parameter values, which will not be known in practice.

In general, the `nlr` package is designed for derivative-based computation and the Nelder–Mead algorithm should only be used in null situations where gradient computations are impossible, for example if it is not possible to compute the gradient at a certain point. Examples of this situation are set out in the rest of this chapter.

Example 8.7   *Carbon dioxide data*
Consider the scaled exponential convex model fitted to the carbon dioxide data in Example 8.5. The OLS estimates are $p_1 = 280.36, p_2 = 32.38$ and $p_3 = 0.01839$. To fit the model using the Nelder–Mead algorithm, we start from initial values $\theta = (p_1, p_2, p_3)$ and three more points are incremented by $\delta = (\delta_1, \delta_2, \delta_3)$. If we run the command as follows:

```
carbon.olsNM <- nlr(
    formula = ScalExp,
    data = crbdt,
    control = nlr.control(
        method = "OLS",
        derivfree = T,
        maxiter = 150
        )
    )
carbon.olsNM$parameters
plot(carbon.olsNM)
```

the result is unexpectedly far from any expected value, $p_1 = 311.198, p_2 = 81.26$, and $p_3 = 0.03166$), and the plot function produces Figure 8.4. The result is incorrect. Note that this example called the `selfStart` function to compute the initial value. Instead, by manually setting $\delta = (150, 15, .007)$, we can get a better result ($p_1 = 280.3578, p_2 = 32.38388, p_3 = 0.01839$) after 178 iterations (see Figure 8.4). Note that the maximum number of iterations must be set sufficiently high, and the convergence tolerance, which will be compared with relative precision, should be very small (10^{-16}).

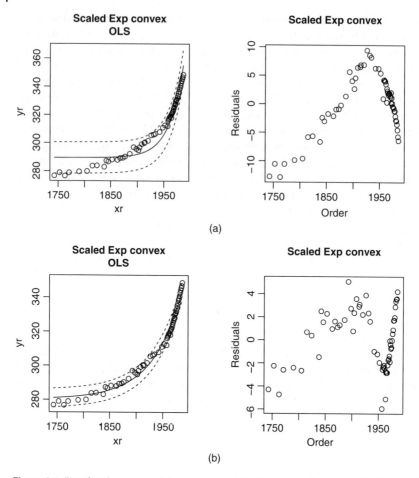

Figure 8.4 Fitted scale exponential convex model for the carbon data, using Nelder–Mead algorithm: (a) using the default δ; (b) using $\delta = c(150, 15, .007)$, and both using selfStart as initial values.

```
carbon.olsNM2 <- nlr(
  formula = ScalExp,
  data = crbdt,
  control = nlr.control(
    method = "OLS",
    derivfree = T,
    maxiter = 500,
    tolerance = 1e-16
  ),
  delta = c(p1=150, p2=15, p3=.007)
)
carbon.olsNM2$parameters
plot(carbon.olsNM2)
```

The `predictionI` function computes the prediction intervals, which can be shown in a graph by the `plot` function. The `ParInfer` function calculates the covariance martrix of the estimated parameter from the equation

$$\widehat{Cov}(\hat{\theta}) = \hat{\sigma}^2 [\dot{\eta}(\hat{\theta}_0)^T \dot{\eta}(\hat{\theta}_0)]^{-1}$$

(given in Chapter 2) and the 0.95% confidence interval for the parameters.

```
> parInfer(carbon.ols)
$covmat
             [,1]          [,2]          [,3]
[1,]   1.707559e+01  6.326218e+02 -1.246155e+06
[2,]   6.326218e+02  7.705323e+21  3.911677e+18
[3,]  -1.246155e+06  3.911677e+18  1.985961e+15


$corrmat
             [,1]          [,2]          [,3]
[1,]   1.000000e+00  1.744058e-09 -0.00676704
[2,]   1.744058e-09  1.000000e+00  0.99995894
[3,]  -6.767040e-03  9.999589e-01  1.00000000


$parstdev
[1]  4.132262e+00  8.777997e+10  4.456413e+07


$CI
           cilow          ciupp
p1   2.980508e+02   3.243455e+02
p2  -2.792835e+11   2.792835e+11
p3  -1.417866e+08   1.417866e+08
```

8.7 Robust Estimators

By default, the `nlr` function estimates the parameters using the MM-estimate. The robust MM-estimate is computed by Algorithm 3.1. The `nl.form` objects that should be prepared in advance are an object for the nonlinear regression function model $f(x_i; \theta)$ assigned, and the robust loss function $\rho(t)$.

Example 8.8 *Methane gas data*
Consider the methane gas data in Table A.2. Four models proposed for this data are given in Equation (3.22) and are saved in the `nlrobj5` variable, accessible from the `nlr` package. The scaled exponential (3.22a) is accessible as `nlobj5[[4]]`, the scaled exponential convex (3.22b) is accessible as `nlobj5[[8]]`, the power model 3.22c is accessible as `nlobj5[[9]]`, and the exponential with intercept (3.22d) is accessible as `nlobj5[[18]]`. The command:

```
mtd <- list(xr = nlr::methane$year,
        yr = nlr::methane$methane)
```

```
st <- list(p1 = 700,
           p2 =.4,
           p3 = 1400,
           p4 = 70)
methane.seMM <- nlr(
  formula = nlrobj5[[4]],
  data = mtd,
  start = st,
  control = nlr.control(tolerance = 1e-16)
)
```

fits the scaled exponential to the methane data, using the Hampel ρ function by default. To compute the OLS estimate for this model, note that the derivative with respect to p_1 is 1 and the rank of the gradient matrix is 3. Therefore, the gradient matrix will be singular. The command below returned a Fault error, with the message "Singular gradient matrix".

```
methane.seOLS <- nlr(
  formula = nlrobj5[[4]],
  data = mtd,
  start = st,
  control = nlr.control(method = "OLS",
                        tolerance = 1e-8)
)
```

This happened because the computation algorithm uses the QR decomposition of the gradient matrix. There are several remedies for this problem, one of which is to use the derivative-free method:

```
methane.seNM <- nlr(
  formula = nlrobj5[[4]],
  data = mtd,
  start = st,
  control = nlr.control(
    method = "OLS",
    derivfree = T,
    tolerance = 1e-16,
    maxiter = 500
  )
)
```

parInfer(methane.seMM) calculates the covariance matrix as in Equation 3.9, and parInfer(methane.seNM) calculates the covariance matrix for OLS. The results are shown in Table 3.1.

The robust least median squares estimate can be performed by setting the control=nlr.control(method="lms") argument. This is a high breakdown point but is not necessarily unique and efficient. In fact, nlr uses least median squares estimates as the starting point when attempting to calculate the MM-estimate. Whenever the initial values are not known by user, this can be a good starting point. Although a genetic algorithm is used by `nlr` to calculate the least median squares estimate, in some situations this might fail to converge. For instance, for the data in Example 8.8, the least median squares fails to converge to an accurate estimate when starting from far-away values such as $p_1 = 280, p_2 = .8, p_3 = 1750$, and $p_4 = 50$, but the MM-estimate converges on the same values as before. In the following example, the OLS, MM and LTS estimates are computed from a far-away initial value and plotted together using the `mplot` function, with which it is possible to plot a list of `nlr` output objects together.

Example 8.9 Starting from a far-away initial value ($p_1 = 280, p_2 = .8, p_3 = 1750, p_4 = 50$) in Example 8.8, the least median squares and MM-estimators can be performed by the following commands:

```
mtd <- list(xr = nlr::methane$year,
            yr = nlr::methane$methane)
st <- list(p1 = 700,
           p2 = .4,
           p3 = 1400,
           p4 = 70)
   methane.seLMS <- nlr(
       formula = nlrobj5[[4]],
       data = mtd,
       start = st,
       control = nlr.control(method = "lms",
                    tolerance = 1e-16)
   )
         # A very far initial value
       st <- list(p1 = 280,
                  p2 = .8,
                  p3 = 1750,
                  p4 = 50)
   methane.seMM2 <- nlr(
     formula = nlrobj5[[4]],
     data = mtd,
     start = st,
     control = nlr.control(tolerance = 1e-16)
     )
   methane.seLMS2 <- nlr(
     formula = nlrobj5[[4]],
     data = mtd,
     start = st,
     control = nlr.control(method = "lms",
                    tolerance = 1e-16)
     )
   methane.seNM2 <- nlr(
       formula = nlrobj5[[4]],
       data = mtd,
       start = st,
```

```
            control = nlr.control(
            method = "OLS",
            derivfree = T,
            tolerance = 1e-16,
            maxiter = 500
            )
)
mtf <- list(NULL)
mtf[[1]] <- methane.seLMS
mtf[[2]] <- methane.seNM2
mtf[[3]] <- methane.seMM2
mtf[[4]] <- methane.seLMS2
mplot(mtf)
```

As can be seen in Figure 8.5, the least median squares estimate starting from a far point is not accurate enough, but compared to far-away initial values it can be a good breakdown point starting value.

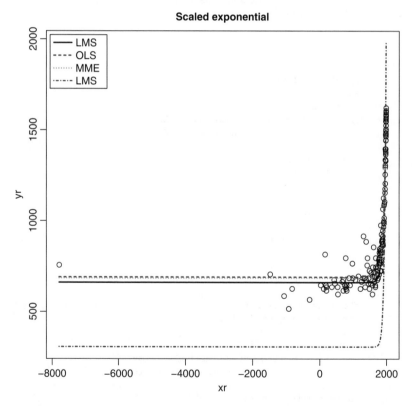

Figure 8.5 Fitted scale exponential convex model for methane gas data using least mean squares. *Source*: Riazoshams and Midi (2013). Reproduced with permission of University Putra Malaysia.

The convergence tolerance `tolerance=1e-16` argument is extremely small and is not recommended for all cases and simulations: it might increase the number of iterations and cause rounding errors in computing the relative precision. For example, the Nelder–Mead method estimates history values using the following command:

```
>methane.seNM$history
...
[185,]      317   603625.6  691.3089  5.687509e-01
     1428.167  75.50851 1.542881e-15
[186,]      319   603625.6  691.3089  5.687509e-01
     1428.167  75.50851 3.857203e-16
[187,]      321   603625.6  691.3089  5.687509e-01
     1428.167  75.50851 4.242923e-15
[188,]      323   603625.6  691.3089  5.687509e-01
     1428.167  75.50851 0.000000e+00
```

This shows that the convergence criterion at the final iteration (188) is zero, so it cannot be reached in all iteration procedures.

8.8 Heteroscedastic Variance Case

Heteroscedastic variance nonlinear regression models were discussed in Chapter 4. The `nlr` package includes all of the methods to fit heteroscedastic models discussed in that chapter. The primary `nl.form` objects required in the `nlr` function are one object for the nonlinear function model f, one object for the robust loss function ρ, and one object for the heterogeneous nonlinear variance function model G (Equation 4.4). The variance function model object can be sent to the `nlr` function via the `varianceform` argument. The `nlr` function automatically considers the heteroscedasticity case if this argument is given. In this case, the default robust multistage estimator (RME) method (Algorithm 4.2) is used for the model fit. Other methods, such as the CME (Algorithm 4.1), RGME (Algorithm 4.4), CLsME (Algorithm 4.3), and weighted M-estimate (WME) (Equation 4.29), can be performed by calling the `control=nlr.control(method="method name")` argument in the `nlr` function.

The four cases shown in Equations (4.2)–(4.5) can be fitted by employing different arguments. The first two general cases (Equations (4.4) and (4.5)) can be performed by calling the `covariance` argument in the `nlr` function, and the latter two cases (Equations (4.4) and (4.5)) can be performed by calling the `varianceform` argument in `nlr`.

8.8.1 Chicken-growth Data Example

Consider the chicken-growth data given in Table A.1. There are several S-shaped nonlinear curves that can be fitted to the data. Riazoshams and Midi

(2009) showed that the logistic model (8.1), with the power function model as a function of the nonlinear model f, can fit the data:

$$y_i = \frac{a}{1 + be^{-cx_i}} \tag{8.1a}$$

$$\sigma_i^2 = \sigma^2 f^\tau(x_i; \theta) \tag{8.1b}$$

$$= \sigma^2 \left[\frac{a}{1 + be^{-cx_i}} \right]^\lambda$$

The logistic model is accessible through the `nlrobj1[[14]]` variable in the `nlr` package, and the variance power model is saved in the `nlrobjvarmdls1[[1]]` variable. To achieve better performance in the iteration procedures, a `selfStart` slot can be defined for the variance model. By taking the logarithm of the power function $v = \sigma^2 t^\lambda$, we have $\log(v) = \log(\sigma^2) + \lambda \log(t)$, so the parameters σ and λ can be found by linear regression. This is implemented in the `selfStart` slot for the `nlrobjvarmdls1[[1]]` object.

Finally, the model fit of the logistic model with the heterogeneous variance power model, using the CME algorithm, can be obtained by:

```
chicken.CME <- nlr(
    formula = nlrobj1[[14]],
    data = b5,
    start = list(p1 = 2300, p2 = 38, p3 = .11),
    tau = list(sg =.09, landa = 2),
    varianceform = nlrobjvarmdls1[[1]],
    control = nlr.control(tolerance =.001, method = "CME")
)
```

where the `tau` argument is the initial value list of the variance model parameter and the `method` argument in the `control` argument is necessary because `nlr` uses robust estimators as the default for many cases. Therefore, to fit the model using RME, we can omit the `method` argument or write it manually to be equal to RME, and specify the robust loss function ρ:

```
chicken.RME <- nlr(
    formula = nlrobj1[[14]],
    data = chdt,
    start = list(p1 = 100, p2 = 42, p3 = .11),
    robustform = "hampel",
    tau = list(sg =.09, landa = 2),
    varianceform = nlrobjvarmdls1[[1]]
)
```

The `chicken.CME` is the `nl.fitt.gn` object, and `chicken.RME` is the `nl.fitt.rgn` object. These include the `hetro` slot, which contains the fitted information of the heterogeneous variance function. For example, the variance parameters can be obtained by:

```
> chicken.RME$parameters
$p1
[1] 2243.547

$p2
[1] 51.88813

$p3
[1] 0.1316549

$sigma
[1] 1.817448
```

and

```
> chicken.RME$hetro$parameters
$sg
[1] 0.0633933

$landa
[1] 2.103011
```

The parameter covariance matrix can be obtained by the `parInfer` method. The covariance matrix of the RME is calculated by the generalized least squares method using Equation 2.21. The heteroscedastic variance model can be fitted by other two methods, CLsME and RGME, using the following commands:

```
chicken.CLSME<- nlr(formula=nlrobj1[[14]], data=chdt,
    start=list(p1=2300,p2=38,p3=.11),tau=list(sg=.09,
                                              landa=2),
    varianceform=nlrobjvarmdls1[[1]],
    control=nlr.control(tolerance=.001,method="CLSME"))
chicken.RGME <- nlr(formula=nlrobj1[[14]],data=chdt,
    start=list(p1=100,p2=42,p3=.11),
    robustform ="hampel",tau=list(sg=.09,landa=2),
    varianceform=nlrobjvarmdls1[[1]],
    control=nlr.control(method="RGME"))
```

The fitted models graph can be drawn using

```
plot(chicken.CME,control=nlr.control(singlePlot=T))
```

where `control=nlr.control(singlePlot=T)` means the fitted curve and residual plots are drawn on separate pages, with the default setting being that they are plotted in two columns. The four models are shown in Figure 8.6, where RME exhibits a narrower prediction interval.

Although the chicken-growth data do not have outliers, the robust methods perform well. To compute the covariance and correlation matrices, standard errors, and confidence intervals for the parameter estimates of the computed robust estimate, the `parInfer` method can be used:

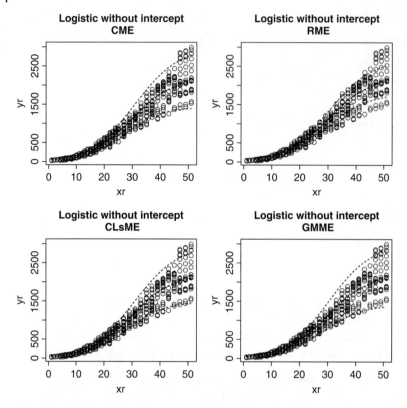

Figure 8.6 Fitted logistic model with the power variance function model for chicken-growth data. *Source*: The Performance of a Robust Multistage Estimator in Nonlinear Regression with Heteroscedastic, Errors, Communications in Statistics – Simulation and Computation 2016. Adapted from Simulation and Computation 2016.

```
> parInfer(chicken.RME)
$covmat
[,1]          [,2]               [,3]
[1,]   2.269297e+00 2.960685e-02 -5.291291e-05
[2,]   2.960685e-02 1.613111e-03  4.701871e-07
[3,]  -5.291291e-05 4.701871e-07  2.793226e-09

$corrmat
[,1]        [,2]        [,3]
[1,]   1.0000000 0.489344 -0.6646039
[2,]   0.4893440 1.000000  0.2215060
[3,]  -0.6646039 0.221506  1.0000000

$parstdev
[1] 1.506419e+00 4.016355e-02 5.285098e-05
```

```
$CI
cilow        ciupp
p1 2238.8893662 2248.2039654
p2   51.7639595   52.0123017
p3    0.1314915    0.1318183
```

Two kinds of outlier might occur for the heteroscedastic variance case: outliers in data and outliers in variance. Sections 6.7and 8.10 provide a discussion of outliers and outlier detection.

8.8.2 National Toxicology Study Program Data

Table A.7 shows mouse kidney data from the National Toxicology Study Program (Bucher, 2007). Lim et al. (2010, 2012) proposed the Hill model, defined as:

$$y_{ij} = \theta_0 + \frac{\theta_1 x_i^{\theta_2}}{\theta_3^{\theta_2} + x_i^{\theta_2}} + \sigma(x_i; \tau)\varepsilon_{ij}, i = 1, \dots, 7, j = 1, \dots, 4 \tag{8.2}$$

to fit the relation between dose concentration (x) and chromium concentration (y) in the blood and kidneys of a mouse. Index j represents the number of replicated design points i, x_i. The authors showed that the heteroscedastic variance can be expressed in a nonlinear form:

$$\sigma(x_i; \tau) = \tau_0 + \frac{\tau_1}{1 + e^{-\tau_2 x_i}}, i = 1, \dots, n \tag{8.3}$$

Unlike the chicken-growth data example (Section 8.1), this variance model is an explicit function of the independent variable. To fit this model using `nlr`, the `data` argument must include the data for the variance function model, with the name similar to the `independent` slot of the variance function `nl.form` object. In the `nlr` package, the Hill model (8.2) is defined in the `nlrobj1[[16]]` variable, and the variance model (8.2) is defined in the `nlrobjvarmdls3[[2]]` variable. The WME approach discussed in Section 4.8 is used by Lim et al. (2010). The WME can be computed by the following commands (Figure 8.7):

```
datalist = list(xr = ntp$dm.k, yr = ntp$cm.k)
datalist[[nlrobjvarmdls3[[2]]$independent]] <- ntp$dm.k
ntpstart = list(p1 = .12,
                p2 = 6,
                p3 = 1,
                p4 = 33)
ntpstarttau = list(tau1 = -.66,
                   tau2 = 2,
                   tau3 =.04)
mouse.WME <- nlr(
  formula = nlrobj1[[16]],
```

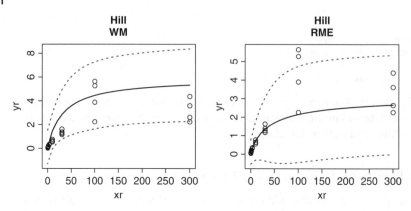

Figure 8.7 Fitted Hill model and Hill variance function model for mouse kidney data. WM Weighted M-estimate, RME Robust Multi Stage Estimate.

```
    data = datalist,
    robustform = "hampel",
    start = ntpstart,
    tau = ntpstarttau,
    varianceform = nlrobjvarmdls3[[2]],
    control = nlr.control(method = "WME")
)
```

in which ntp is the data set stored in nlr package. The datalist is the list of variables, including dependent variables and independent variables, and with repeated independent variables with same name as the nlrobjvarmdls3[[2]]$independent character name. The method="WME" specifies that the WME method should be employed. Note that the nlr function identifies that the heteroscedasticity case is required to be fitted from the varianceform argument. If the control argument is not given, the default RME method will be used, as in the following command:

```
        mouse.RME <- nlr(formula=nlrobj1[[16]],
            data=datalist,
            robustform ="hampel",
            start=ntpstart,
            tau=ntpstarttau,
            varianceform=nlrobjvarmdls3[[2]])
```

8.9 Autocorrelated Errors

The autocorrelated errors case in nonlinear regression models is discussed in Chapter 5. The correlation argument in the nlr function accepts the correlation structure object (corStruct) from the nlme R-package or calls the user-defined autocorrelation function. The autocorrelated cases that are available in `corStruct` are "corAR1", "corARMA", "corCAR1", "corCompSymm", "corExp", "corGaus", "corLin", "corRatio", "corSpher", and "corSymm". In the `nlr` package at the time of writing, only "corAR1", for autoregressive order 1, and "corARMA", for the autoregressive moving average process, with arbitrary orders for the autoregressive and moving average components, and capability for defining the user autocorelated structure, were operational. Other cases are under development. The `correlation` argument is a list of commands as follows:

`StructName="NAN"`: String name of the correlation structure. This can be names of `corrStruct` from the nlme package, including "corAR1", "corARMA", "corCAR1", "corCompSymm", "corExp", "corGaus", "corLin", "corRatio", "corSpher", and "corSymm". In this case the appropriate commands to be submitted to `nlem`, such as order and parameters, must be included as the list items. The manual definition `StructName="manual"` means the user will give a personally defined autocorrelation function. In this case, the user-defined function will be attached in the list via the `manualcorr` name or function body.

`manualcorr=NULL`: Manual name or function body of a user-defined autocorrelation function. The user-defined autocorrelation function must include an entry argument named `ri`, for the residuals, the output of the list with names `tm` as the fitted time series object, and `vmat` as the autocorrelation matrix V. Later this will be used in QR decomposition and the generalized fit.

To fit a nonlinear regression model with autocorrelated errors, the `correlation` has to be set to a `corStruct` object. For example, consider the trademark data for Iran given in Table A.6. The data include an x-direction outlier. Since the data were collected over time, it is possible that their errors are correlated. To identify the autocorrelation structure, and since there are outliers, first the model has to be fitted using a robust method or a classical method with removal of outliers. Then we compute the residuals and try to detect the autocorrelation structure. Finally, we can fit the model using the autocorrelation structure using the robust two-stage estimate (Algorithm 5.2) or the two-stage estimate (Algorithm 5.1) if we have a clean data set.

Example 8.10   Consider the Iran trademark data given in Table A.6 (see Section 5.5). The data represent 47 years of observations of trademark applications (direct residents) from 1960 to 2007, reported by data.worldbank.org in 2013. As discussed in Section 5.5, first we have to choose a nonlinear model. According to the increasing exponential trend of the data, the four models in Equation 3.22 might be appropriate for this data set. An alternative is to take the logarithms of both variables and try to fit a linear regression model, but we will not take that approach. The following command fits the four models, including the scale exponential model (3.22a) to the Iran trademark data using robust and OLS methods for comparison. As shown in Section 5.5, the correlation structure follows the $ARIMA(1, 0, 1)(0, 0, 1)7$ model. To compute the robust fit using the RTS estimate, we have to define a function for computing the auto-correlation function (Equation 5.14), then send it to the `nlr` package using the `correlation (= list(StructName="manual", manual-corr="arima_r1010017")` argument, where `StructName="manual"` means a user-defined function computed by the autocorrelation function and `manualcorr = "arima_r1010017"` represents the function that performs the computation. The commands for the procedures explained in Section 5.5 are as follows:

```
library(nlr)
tadr <- nlr::trade.ir
p1 <- min(tadr[, 2])#8.06e+10
p2 <- 1e11
p3 <- 1970
p4 = 6
irstrt <- list(p1 = p1,
               p2 = p2,
               p3 = p3,
               p4 = p4)
fitt.irtrade <- list(NULL)
# Model 1:  yr = p_1 + p_2 * e^{(xr - p_3)/p_4}
# Use SelfStart defined in object
irtadr.MM <- nlr(
  nlrobj5[[4]],
  data = list(xr = tadr[, 1],
              yr = tadr[, 2]),
  control = nlr.control(
    trace = T,
    tolerance =
      1e-6,
    maxiter = 2000
  )
)


ir.tadrOLS1 <- nlr(
  nlrobj5[[4]],
  data = list(xr = tadr[, 1],
              yr = tadr[, 2]),
  control = nlr.control(
    trace = T,
    method = "OLS",
    derivfree = T,
    tolerance = 1e-6,
    maxiter = 2000
  )
)
```

```
fitt.irtrade[[1]] <- ir.tadrOLS1
fitt.irtrade[[2]] <- irtadr.MM
# Model 2: yr = p_1 + e^{-(p_2 - p_3 * xr)}
fitt.irtrade[[3]] <- nlr(
  nlrobj5[[8]],
  data = list(xr = tadr[, 1], yr = tadr[, 2]),
  control = nlr.control(
    trace = T,
    method = "OLS",
    derivfree = F,
    tolerance = 1e-6,
    maxiter = 3000
  )
)
fitt.irtrade[[4]] <- nlr(
  nlrobj5[[8]],
  data = list(xr =
                 tadr[, 1], yr = tadr[, 2]),
  control = nlr.control(
    trace = T,
    derivfree = F,
    tolerance = 1e-6,
    maxiter = 2000
  )
)
#' Model 3:
#' $yr = 1/p_ - p_2 * p_3^xr$
#' Initial values calculate by selfStart
fitt.irtrade[[5]] <- nlr(
  nlrobj5[[9]],
  data = list(xr = tadr[, 1],
              yr = tadr[, 2]),
  control = nlr.control(
    trace = T,
    method = "OLS",
    derivfree = T,
    tolerance = 1e-4,
    maxiter = 4000
  )
)
fitt.irtrade[[6]] <- nlr(
  nlrobj5[[9]],
  data = list(xr = tadr[, 1],
              yr = tadr[, 2]),
  #start=chstart3,
  control = nlr.control(
    trace = T,
    derivfree = F,
    tolerance = 1e-4,
    singularCase = 2,
    maxiter = 8000
  )
)
```

```
#' Model 4:
#' $yr = p_1 + e^{(xr - p_2)/p_3}$
#'
fitt.irtrade[[7]] <- nlr(
  nlrobj5[[18]],
  data = list(xr =
                  tadr[, 1][-44], yr = tadr[, 2][-44]),
  control = nlr.control(
    trace = T,
    method = "OLS",
    derivfree = F,
    tolerance = 1e-4,
    maxiter = 4000
  )
)
fitt.irtrade[[8]] <- nlr(
  nlrobj5[[18]],
  data = list(xr =
                  tadr[, 1], yr = tadr[, 2]),
  control = nlr.control(
    trace = T,
    derivfree = F,
    tolerance = 1e-4,
    singularCase = 1,
    maxiter = 8000
  )
)
#'  Inferences, robust results
#'  Model 1
fitt.irtrade[[2]]$form$name
fitt.irtrade[[2]]$htheta
fitt.irtrade[[2]]$parameters
parInfer(fitt.irtrade[[2]])
fitt.irtrade[[2]]$correlation
nrow(fitt.irtrade[[2]]$history)

#' model 2
fitt.irtrade[[4]]$form$name
fitt.irtrade[[4]]$htheta
fitt.irtrade[[4]]$parameters
parInfer(fitt.irtrade[[4]])
fitt.irtrade[[4]]$correlation
nrow(fitt.irtrade[[4]]$history)

#' model 3
fitt.irtrade[[6]]$form$name
fitt.irtrade[[6]]$htheta
fitt.irtrade[[6]]$parameters
parInfer(fitt.irtrade[[6]])
fitt.irtrade[[6]]$correlation
nrow(fitt.irtrade[[6]]$history)
```

```
#' model 4, selected.
fitt.irtrade[[8]]$form$name
fitt.irtrade[[8]]$htheta
fitt.irtrade[[8]]$parameters
parInfer(fitt.irtrade[[8]])
fitt.irtrade[[8]]$correlation
nrow(fitt.irtrade[[8]]$history)
#' Identify autocorrelation for model 4
#' Using robust
ri.MM1 <- residuals(fitt.irtrade[[8]])
ri.ols1 <- residuals(fitt.irtrade[[7]])
plot(fitt.irtrade[[8]])
plot(fitt.irtrade[[7]])
mlist <- list(fitt.irtrade[[8]], fitt.irtrade[[7]])
mplot(mlist)
par(mfrow = c(3, 2))
acf(ri.MM1)
pacf(ri.MM1)
acf(ri.ols1)
pacf(ri.ols1)
robacf(ri.MM1)
robacf(ri.ols1)
# remove outlier
ri.MM2 <- ri.MM1[-44]
ri.ols2 <- ri.ols1[-44]
par(mfrow = c(3, 2))
acf(ri.MM2)
pacf(ri.MM2)
acf(ri.ols2)
pacf(ri.ols2)
robacf(ri.MM2)
robacf(ri.ols2)

# removed outlier
fitt.irtrade[[9]] <- nlr(
  nlrobj5[[18]],
  data = list(xr =
                tadr[, 1][-44], yr = tadr[, 2][-44]),
  control = nlr.control(
    trace = T,
    method = "OLS",
    derivfree = F,
    tolerance = 1e-4,
    maxiter = 4000
  )
)
fitt.irtrade[[10]] <- nlr(
  nlrobj5[[18]],
  data = list(xr = tadr[, 1][-44], yr = tadr[, 2][-44]),
  control = nlr.control(
    trace = T,
```

```
      derivfree = F,
      tolerance = 1e-4,
      singularCase = 1,
      maxiter = 8000
  )
)
ri.MM3 <- residuals(fitt.irtrade[[10]])
ri.ols3 <- residuals(fitt.irtrade[[9]])
par(mfrow = c(3, 2))
acf(ri.MM3)
pacf(ri.MM3)
acf(ri.ols3)
pacf(ri.ols3)
robacf(ri.MM3)
robacf(ri.ols3)

#' ****************************
#' Forecast the outlier point
#'    replace and fit model again
#'

xr = tadr[, 1]
yr = tadr[, 2]
newdata <- list(xr = tadr[, 1][44], yr = tadr[, 2][44])
yr[44] <- predict(fitt.irtrade[[8]], newdata = newdata)
a1 <- nlr(
  nlrobj5[[18]],
  data = list(xr = xr, yr = yr),
  control = nlr.control(
    trace = T,
    method = "OLS",
    derivfree = F,
    tolerance = 1e-4,
    maxiter = 4000
  )
)
fitt.irtrade[[11]] <- a1
a1 <- nlr(
  nlrobj5[[18]],
  data = list(xr = xr, yr = yr),
  control = nlr.control(
    trace = T,
    derivfree = F,
    tolerance =
      1e-4,
    singularCase = 1,
    maxiter = 8000
  )
)
fitt.irtrade[[12]] <- a1
# identify autocorrelation
```

```
ri.MM4 <- residuals(fitt.irtrade[[12]])
ri.ols4 <- residuals(fitt.irtrade[[11]])
par(mfrow = c(3, 2))
acf(ri.MM4)
pacf(ri.MM4)
acf(ri.ols4)
pacf(ri.ols4)
robacf(ri.MM4)
robacf(ri.ols4)

#. *******************************
#' Two stage estimate
#'  ARMA (1,0,1)(0,0,1)7
#'
library(forecast)
ri <- residuals(fitt.irtrade[[12]])
tsdisplay(ri)
cr <- arimax(ri,
             order = c(1, 0, 1),
             seasonal = list(order =
                             c(1, 0, 1), period = 7))
summary(cr)
rid <- diff(ri)
acf(rid)
pacf(rid)
auto.arima(ri)
tsdisplay(diff(ri, 7))
tsdisplay(diff(diff(ri, 7), 7))
cr <- arimax(ri,
             order = c(1, 0, 1),
             seasonal = list(order =
                             c(0, 0, 1), period = 7))
fit1 <- arimax(ri, order = c(1, 0, 1))
fit2 <- arimax(ri,
               order = c(1, 0, 1),
               seasonal = list(order =
                               c(0, 0, 1), period = 7))
fit3 <- arima(ri,
              order = c(1, 0, 1),
              seasonal = list(order =
                              c(0, 0, 1), period = 7))
fit1
fit2
#' ***********************
#' *  Comparing other models
#' *
arimax(ri,
       order = c(1, 0, 2),
       seasonal = list(order =
                       c(0, 0, 8), period = 7))
arima(ri,
```

```
       order = c(1, 0, 1),
       seasonal = list(order =
                       c(0, 0, 1), period = 7))

#' **********************************************
#'   User defined autocorrelation function
#'   For ARIMA(1,0,1)(0,0,1)7
#'   Derived by Hossein Riazoshams
#'   Feb, 2017
#'   Robust Nonlinear Regression
#'   With Applications using R
#'   Hossein Riazoshams, Habshah Midi,
#'   Gebrenegus Ghilagaber
#'
arima_r1010017 <- function(ri) {
  tm <- arimax(
    ri,
    order = c(1, 0, 1),
    seasonal = list(order =
                    c(0, 0, 1), period =
                    7),
    include.mean = FALSE
  )
  phi <- as.numeric(tm$coef[1])
  theta <- as.numeric(tm$coef[2])
  tseasonal <- as.numeric(tm$coef[3])
  n <- length(ri)
  sigma2a <- tm$sigma2
  zeta0 <- (1 + tseasonal ^ 2) * sigma2a
  gam <- 0
  gam <- ((1 + theta ^ 2 - 2 * theta * phi) / (1 - phi ^ 2)) *
zeta0
  gam[2] <- (((phi - theta) * (1 - phi * theta)) / (1 - phi ^ 2))
  * zeta0
  for (k in 2:7) {
    gam[k + 1] <- phi ^ (k - 1) * gam[2]
  }
  gam[9] <- phi ^ 7 * gam[2] + ((tseasonal * (1 - theta ^ 2)) /
(phi))
  * sigma2a for (k in 9:(n - 1)) {
    gam[k + 1] <- phi ^ (k - 8) * gam[9]
  }
  rho <- gam / gam[1]
  vmat <- matrix(rep(0, n * n), nrow = n)
  for (i in 1:(n - 1)) {
    rho2 <- 0
    rho2 <- rho[i:1]
    rho2[(i + 1):n] <- rho[2:(n - i + 1)]
    vmat[i,] <- rho2
  }
  vmat[n,] <- rho[n:1]
```

```
vmat[vmat < 1e-8] = 0
return(list(tm = tm, vmat = vmat))
}

#' Two Stage Estimate
xr = tadr[, 1]
yr = tadr[, 2]
a1 <- nlr(
  nlrobj5[[18]],
  data = list(xr = xr, yr = yr),
  control = nlr.control(
    trace = T,
    derivfree = F,
    tolerance = 1e-8,
    singularCase = 1,
    maxiter = 8000
  ),
  correlation = list(StructName = "manual",
                     manualcorr = "arima_r1010017")
)     # deriv MM, selfstart
fitt.irtrade[[13]] <- a1
fitt.irtrade[[13]]$form$name
fitt.irtrade[[13]]$htheta
fitt.irtrade[[13]]$parameters
parInfer(fitt.irtrade[[13]])
fitt.irtrade[[13]]$correlation
```

8.10   Outlier Detection

To detect outliers, the `nlr` package uses some statistical measures. Ria-zoshams et al. (2011) used six statistical measures and showed that studentized residuals and the Cook distance can detect outliers only when combined with robust methods. Using these results, two functions were embedded in the `nlr` package to detect outliers. After fitting a model by calling `atypicals`, `atypicals.deleted` functions are used to calculate the statistical measures. The `atypicals` calculated are "vmat", "studres", "cook", "mahd.v", "mahd.dt", "mahd.xs", "hadi", "potmah", "mvedta", and "mvex" based on the full data set, and "jl.vmat", "jl.studres", "jl.cook", and "jl.hadi" based on Jacobian leverages. In addition, `atypicals.deleted` calculates "yihat", "delstud", "dffits", "atk", and "dfbetas" based on deleting one point at a time. In nonlinear regression we do not recommend that points are deleted because the model curve can be affected.

Example 8.11   *Lakes data*
Consider the lakes data example in Section 6.6. The nonlinear model given by Equation 6.23 is created as an `nl.form` object saved in an `nlrobj5[[2]]`

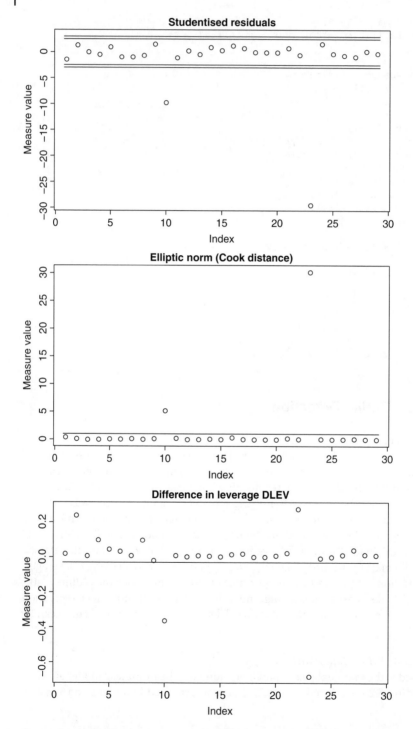

Figure 8.8 Outlier detection measures for lakes data: studentized residual, Cook distance, and *DLEV*.

variable. The following commands fit the model using the MM-estimate. It can been seen that the initial values are effective in achieving a high breakdown estimate, so the LMS estimate is used for the initial value. The maximum number of iterations is set to 5000 and convergence is slowly achieved, after 2328 iterations. The outlier detection measures discussed in Chapter 6 can be computed by atypicals. The *DLEV* measure is computed using the dlev function defined for a nl.fitt.rob object, and creates a result in the nl.robmeas object, including the measure value and cut-off point. The plot function adjusted for the nl.robmeas object produces Figure 8.8. The program codes include computation of the tangential and Jacobian leverages, and their differences are discussed in Section 6.6. The graph of these leverages is shown in Figure 8.9.

```
LakesFitt<- nlr(nlrobj5[[2]],
  data=list(xr1=Lakes$nin,xr2=Lakes$tw,yr=Lakes$tn),
  control=nlr.control(tolerance=1e-10,
          initials="lms",
          maxiter = 5000))
hatmat <- hat(LakesFitt)
jaclevmat <- JacobianLeverage(LakesFitt)
ad1 <- diag(hatmat)
ad2 <- diag(jaclevmat)
dlevdiff <- ad2 - ad1
dlevrelat <- dlevdiff
cutofpoint <- median(dlevdiff) - 3 * mad(dlevdiff)
plot(
  ad2,
  ylim = c(0,.7),
  type = "b",
  main = "Leverages",
  xlab = "Data order",
  ylab = "values"
)
legend(
  1,
  .7,
  c("Robust Jacobian leverage",
      "Tangent Jacobian leverage"),
  pch = c(1, 2),
  cex =.7
)
mtext("Lakes Data, MM-estimate")
lines(ad1, type = "b", pch = 2)
plot(
  ad2,
  ylim = c(0,.7),
  type = "b",
  main = "Leverages",
  xlab = "Data order",
  ylab = "values"
)
```

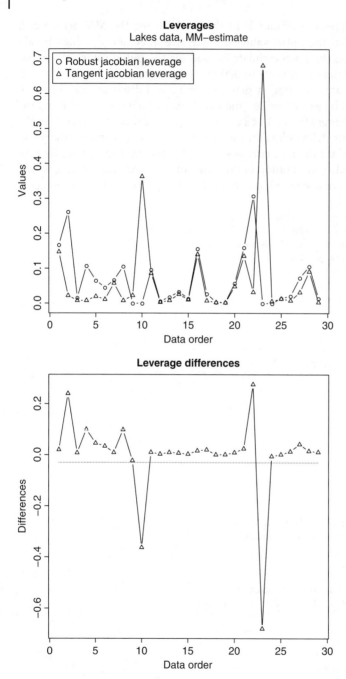

Figure 8.9 Tangential plane leverage, robust Jacobian leverage and their differences for lakes data.

```
legend(
  1,
  .7,
  c("Robust Jacobian leverage", "Tangent Jacobian leverage"),
  pch = c(1, 2),
  cex = .7
)
mtext("Lakes Data, MM-estimate")
lines(ad1, type = "b", pch = 2)
```

Example 8.12 Consider the simulated example for the heteroscedastic variance error given in Section 6.7. The simulation, the fitted model for the simulated data, outlier detection (by atypicals function), and plot commands are as follows:

```
library(nlr)
exactteta <- list(p1 = 2570, p2 = 41, p3 = 0.1)
exactteta2 <- list(p1 = 2570,
                   p2 = 41,
                   p3 = 0.08,
                   sigma = .08)
exacttau <- list(sg =.08, landa = 1.06)
exactteta <- list(p1 = 2213, p2 = 52, p3 = 0.1335976)
exacttau <- list(sg = sqrt(.006189322), landa = 2.225973)
dt.HetroSim <- NULL
a2 <- as.integer(seq(3, 51, length = 10))
a3 <- NULL
for (i in 1:length(a2))
  a3 <- c(a3,a2[i],a2[i],a2[i],a2[i],a2[i],a2[i],
    a2[i],a2[i],a2[i],a2[i],a2[i],a2[i],
    a2[i],a2[i],a2[i],a2[i],a2[i],a2[i],a2[i]
  )
n <- length(a3)
datalist <- list(xr = a3,yr=a2)
datalist[names(exactteta)] <- exactteta
b3 <- eval(nlrobj1[[14]], datalist)     ### f(x)
b3 <- as.numeric(b3$predictor)
sdvmodel <- exacttau$sg ^ 2 * b3 ^ exacttau$landa
sdvmodel <- sqrt(sdvmodel)
fmodel <- sdvmodel  * rnorm(length(a3), mean = 0, sd = 1)
b4 <- b3 + fmodel
#' *****************
#'  Simulate Case 1
dt <- list(xr = a3, yr = b4)
nrp0 <- nonrepl(list(x = dt$xr, y = dt$yr))
z <- zvalues(dt$yr, nrp0$ni, nrp0$xo)
z2 <- rzvalues(dt$yr, nrp0$ni, nrp0$xo)
dt.HetroSim[[1]] <- list(xr = dt$xr, yr = dt$yr)
m <- matrix(c(1, 2, 3),
            nrow = 3,
            ncol = 1,
            byrow = TRUE)
```

```
layout(mat = m, heights = c(0.33, 0.33, .33))
#layout.show(3)
par(mar = c(0, 4, 1.3, 4))
plot(
  dt$xr,
  dt$yr,
  xlab = "x",
  ylab = "y",
  xaxt = 'n',
  yaxt = 'n',
  main = "(a)"
)par(mar = c(0, 4, 0, 4))
plot(
  dt$xr[nrp0$xm],
  z[nrp0$xm],
  xlab = "x",
  ylab =
    "z values",
  xaxt = 'n',
  yaxt = 'n',
  type = "b"
)
par(mar = c(2, 4, 0, 4))
plot(
  dt$xr[nrp0$xm],
  z[nrp0$xm],
  xlab = "x",
  ylab =
    "z values",
  xaxt = 'n',
  yaxt = 'n',
  type = "b"
)
axis(1,
     at = seq(0, 50, 5),
     cex = 1,
     tck = -.08)
#' *****************
#'  Simulate Case 2
y2 <- b4
y2[61:80] = y2[61:80] + 2000
nrp <- nonrepl(list(x = dt$xr, y = y2))
z <- zvalues(y2, nrp$ni, nrp$xo)
z2 <- rzvalues(y2, nrp$ni, nrp$xo)
dt.HetroSim[[2]] <- list(xr = dt$xr, yr = y2)
m <- matrix(c(1, 2, 3),
              nrow = 3,
              ncol = 1,
              byrow = TRUE)
layout(mat = m, heights = c(0.33, 0.33, .33))
par(mar = c(0, 4, 1.3, 4))
```

```
plot(
  dt$xr,
  y2,
  main = "(b)",
  xlab = "x",
  ylab = "y",
  xaxt = 'n',
  yaxt = 'n'
)par(mar = c(0, 4, 0, 4))
plot(
  dt$xr[nrp0$xm],
  z[nrp$xm],
  xlab = "x",
  ylab =
    "z values",
  xaxt = 'n',
  yaxt = 'n',
  type = "b"
)
par(mar = c(2, 4, 0, 4))
plot(
  dt$xr[nrp0$xm],
  z2[nrp$xm],
  xlab = "x",
  ylab =
    "Robust z values",
  xaxt = 'n',
  yaxt = 'n',
  type = "b"
)
axis(1,
     at = seq(0, 50, 5),
     cex = 1,
     tck = -.08)
#' ****************
#'  Simulate Case 3
y1 <- y2 <- b4
t0 <- y2[181:200]
t0
t1 <- mean(t0)
t <- t0 / 12
t <- t + (11 / 12) * t1
y2[181:200] = t
sd(t) ^ 2
var(t)
mad(t) ^ 2
nrp <- nonrepl(list(x = dt$xr, y = y2))
z <- zvalues(y2, nrp$ni, nrp$xo)
z2 <- rzvalues(y2, nrp$ni, nrp$xo)
dt.HetroSim[[3]] <- list(xr = dt$xr, yr = y2)
m <- matrix(c(1, 2, 3),
```

```
                    nrow = 3,
                    ncol = 1,
                    byrow = TRUE)
    layout(mat = m, heights = c(0.33, 0.33, .33))
    par(mar = c(0, 4, 1.3, 4))
    plot(
      dt$xr,
      y2,
      main = "(c)",
      ylab = "y",
      xaxt = 'n',
      yaxt = 'n'
    )
    par(mar = c(0, 4, 0, 4))
    plot(
      dt$xr[nrp$xm],
      z[nrp$xm],
      xlab = "x",
      ylab = "z values",
      xaxt = 'n',
      yaxt = 'n',
      type = "b"
    )
    par(mar = c(2, 4, 0, 4))
    plot(
      dt$xr[nrp$xm],
      z2[nrp$xm],
      xlab = "x",
      ylab =
        "Robust z values",
      xaxt = 'n',
      yaxt = 'n',
      type = "b"
    )
    axis(1,
         at = seq(0, 50, 5),
         cex = 1,
         tck = -.08)


    #' ************************
    #'   Fit model
    fit.HetroSim <- list(NULL)
    oi = 0
    for (i in 1:length(dt.HetroSim)) {
      print(i)
      plot(dt.HetroSim[[i]]$xr, dt.HetroSim[[i]]$yr)
      oi <- oi + 1
      print(oi)
      fit.HetroSim[[oi]] <-
        nlr(
          formula = nlrobj1[[14]],
```

```
      data = dt.HetroSim[[i]],
      start = list(p1 = 2300, p2 = 42, p3 = .11),
      robustform = "hampel",
      tau = list(sg = .09, landa = 2),
      varianceform = nlrobjvarmdls1[[1]],
      control = nlr.control(initials = "lms", method = "CME")
    )   # non robust cme
  oi <- oi + 1
  fit.HetroSim[[oi]] <-
    nlr(
      formula = nlrobj1[[14]],
      data = dt.HetroSim[[i]],
      start = list(p1 = 2300, p2 = 42, p3 =.11),
      robustform = "hampel",
      tau = list(sg =.09, landa = 2),
      varianceform = nlrobjvarmdls1[[1]],
      control = nlr.control(initials = "lms", method = "RME")
    )   # robust rme
}
# - - - - - - - - - outlier detection
outmeas.HetroSim <- NULL
for (oi in 1:length(fit.HetroSim)) {
  print(oi)
  outmeas.HetroSim[[oi]] <- list(NULL)
  a1 = atypicals(fit.HetroSim[[oi]])
  outmeas.HetroSim[[oi]][[1]] <- a1
  for (ja in a1) {
    if (class(ja) == "nl.robmeas")
      plot(ja)
  }
}
a1 = dlev(fit.HetroSim[[4]])
plot(a1)
outmeas.HetroSim[[oi]][[1]] <- a1
par(mfrow = c(2, 2))
plot(outmeas.HetroSim[[2]][[1]][[2]])
plot(outmeas.HetroSim[[2]][[1]][[3]])
plot(outmeas.HetroSim[[4]][[1]][[2]])
plot(outmeas.HetroSim[[4]][[1]][[3]])
plot(outmeas.HetroSim[[6]][[1]][[2]])
plot(outmeas.HetroSim[[6]][[1]])
```

8.11   Initial Values and Self-start

In almost all procedures, a nonlinear regression fit requires initial values for parameter estimates. Among all the iteration options, this is the most critical parameter. At the starting point, the parameter values are not only unknown, but there might be no idea about their possible values. Far away initial data values might be outside the range of parameters; even those inside the range but far away might create a computation error or, more critically, might mislead the computation to a divergent area, so that the iterations do not converge overall. In this case, the user might be misled into thinking that that the model was wrong for the data.

To resolve the initial value problem, `nlr` provides some tools. There are a few choices, but it is important to note that even the best option, using `selfStart`, might not work in some examples.

- `selfStart` slot: The initial values can be computed from the `selfStart` defined function in the `nl.form` object. See Section 8.2.1.
- `initials` argument: The `initials` argument in `nlr.control` represents a computation method for estimating initial values. It can be a string equal to one of the values ("manual", "lms", "OLS", "quantile"). "lms", the least median squares (LMS) estimate, is not unique and efficient, but is a high breakdown point. "OLS" computes ordinary least squares as initial values. "quantile" computes quantile regression by the `nlrq` package. "manual" is an optional value that gets an initial value from the `start` argument.
- `par` slot: The `par` slot from the defined `nl.form` is applied if no option is given by the user. It is an overall value that will not work for all cases and thus it is not recommended and should only be applied if there is no other choice.
- Other estimates: Other estimates, like MLE, can be used as starting points, and the result can then be used as an initial value for the entry for complicated models. This approach can be called by the `method = "MLE"` argument of `nlr.control`.
- Derivative-free methods: Deriviative-free computation can be called by the `derivfree = T` argument of `nlr.control`. Derivative-free methods are sensitive to initial values, but since they do not compute the Hessian and gradient, they are computationally less expensive. If they give slightly closer parameter values these can be used as initial values for derivative-based calculations.
- `plotinitial` function: This function draws the curve for an `nl.form` object based on some data and given initial values or a `selfStart` slot, but only for a two-dimensional graph (see Section 8.2.1).
- Change location and re-scale data: Sometimes data scale and model curvature are effective in computation. Re-scaling data, transforming or modifying the model, can be useful (see Example 8.14.)

It is should be noted that in `nlr`, if a method is not converging, another method will be applied. Seem, for example, the optimization techniques explained in Chapter 7. Nonconvergence might be due to the initial values, so if the output result shows nonconvergence the user can change the initial values. The output result of `fit` includes a slot called `Fault`, which will identify convergence problems.

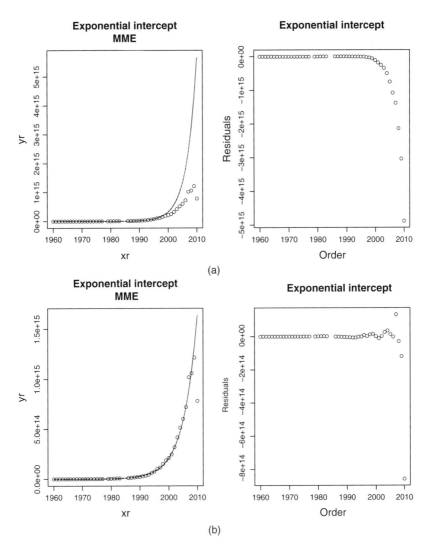

Figure 8.10 Fitted exponential with intercept for Iran net money data using: (a) LMS and (b) selfStart as the initial value.

Example 8.13 As an example of a possible computation problem, consider the Iran net money data given in Table A.5. The code below tries to fit the model using LMS as the initial value but produces a nonconvergent iteration. Figure 8.10a shows the fitted curve. Figure 8.10b shows the correct fit, obtained using the selfStart slot.

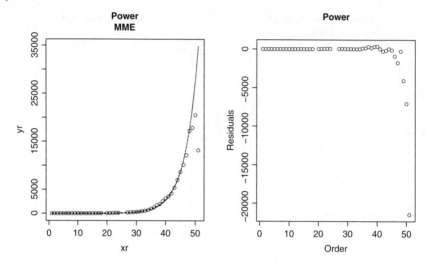

Figure 8.11 Fitted power model for Iran net money data.

```
irandt <- nlr::net.ir
netfit <- nlr(
  nlrobj5[[18]],
  data = list(xr = irandt[, 1],
              yr = irandt[, 2]),
  control = nlr.control(
    tolerance = 1e-8,
    trace =
      T,
    maxiter = 5000,
    initials = "lms",
    singularCase = 2
  )
) # default robust MM
netfit_S <- nlr(
  nlrobj5[[18]],
  data = list(xr =
                irandt[, 1], yr = irandt[, 2]),
  control = nlr.control(
    tolerance = 1e-8,
    trace = T,
    maxiter = 1000
  )
)#default MM
plot(netfit)
plot(netfit_S)
```

Example 8.14   Consider the Iran net money data given in Table A.5. Many unsuccessful attempts were made to fit the power model (Equation 3.22c) given by

$$y_i = \frac{1}{p_1} - p_2 \cdot p_3^{x_i}$$

Using `selfStart` stored in the `nlrobj5[[9]]` object, derivative-free, LMS initial values, OLS estimate, "quantile" regression as the initial value, and many other options were unsuccessful. To remedy the problem, note that the power part $p_3^{x_i}$ for large values of x is very large, especially for the Iran net money data with large x will be overflow from memory. The y values are also large, although these are not directly used in formula, only in the residual computation. Changing the location of x did not work. Thus, we changed the location of x and re-scaled the y data. The fit using the following command was then successful in estimating the model. The fitted curve is shown in Figure 8.11.

```
irandt <- nlr::net.ir
netfit_t <- nlr(
  nlrobj5[[9]],
  data = list(xr = irandt[, 1] - 1959,
              yr = irandt[, 2] / min(irandt[, 2])),
  #start=chstart4,
  control = nlr.control(trace = T, singularCase = 2)
)
plot(netfit_t)
```

9

Robust Nonlinear Regression in R

The `nlr` package discussed in Chapter 8 was produced to deal with the methods discussed in this book. There are another packages in the R language comprehensive archive for robust nonlinear regression. For example, the `nlrq` function in the `nlrq` package developed by Koenker and Park (1996) for quantile regression, the `nlrob` function in the `robustbase` package for M-estimates using iterated reweighted least squares. These tools are explained in current chapter and will be compared with `nlr`, as presented in Chapter 8.

For comparison of the packages, a simulation study will be shown, because the exact values are known and the biases can therefore be computed. Then one easy and one complicated example will be illustrated.

The `nlrq` function from the `nlrq` package fits a nonlinear regression model by quantile regression. It was proposed by Koenker and Park (1996). It is a generalization of the least absolute (l_1 norm) estimate shown in Equation (3.3).

The `nlrob` function in the `robustbase` package fits a nonlinear regression by iteratively reweighted least squares.

9.1 Lakes Data Examples

Consider the lakes data in Example 8.11. The MM-estimate fitted by the `nlr` function is shown in the example. The quantile regression, least median squares (LMS), and ordinary least squares (OLS) estimates are shown at in the code below. The parameter estimates for the four methods are shown in Table 9.1. As the initial value, the LMS figure that is the output from the `lakes_lms` variable is employed for all of the methods. This causes the estimators to be in a similar situation, and their iterations to be fairly robust. From the table, only the MM-estimate computed by `nlr` package is close to being robust. The estimates from `nlrq` and `nlrob` are close to the OLS estimate computed by the `nlr` and `nls` functions. It is worth mentioning that the OLS estimates by the `nlr` and `nls` functions are similar, which makes `nlr` give results similar to standard software packages.

Robust Nonlinear Regression: with Applications using R, First Edition.
Hossein Riazoshams, Habshah Midi, and Gebrenegus Ghilagaber.
© 2019 John Wiley & Sons Ltd. Published 2019 by John Wiley & Sons Ltd.
Companion website: www.wiley.com/go/riazoshams/robustnonlinearregression

Table 9.1 Parameter estimates for lakes data, using the `nlr`, `nlrq`, `nlrob`, and `nls` functions.

| Parameters | Initial values LMS | | Initial values: $p_1=1, p_2=1$ | |
	p_1	p_2	p_1	p_2
`nlr-lms`			0.9644752	0.9373101
`nlr-OLS`	4.801404	1.386543	4.801404	1.267725
`nlr-MM`	0.7647222	0.2975665	1.902971	0.8448515
`nlrq`	3.838413	1.309658	3.835179	1.310451
`nlrob-MM`	4.904799	1.362898	4.9048	1.362901
`nls-OLS`	4.801	1.387	4.801	1.387

Initial values are least median squares, and manually fixed values ($p_1=1, p_2=1$)

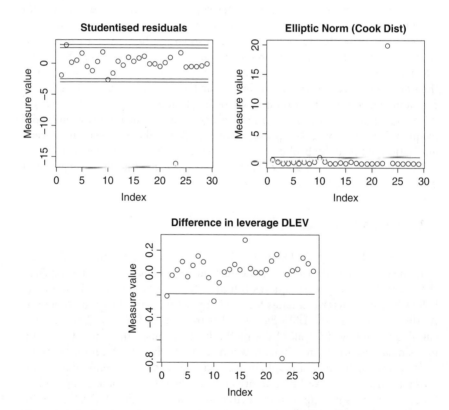

Figure 9.1 Outlier detection measures based on manual initial values ($p_1=1, p_2=1$).

Although `nlrob` is more robust, it is not stable, and it is sensitive to initial values. As shown in Figure 8.11, it is still more robust. It can be seen from Figure 9.1 that if we calculate outlier measures for the estimated model, initial values ($p_1 = 1, p_2 = 1$) give an incorrect outlier detection result for DLEV, the studentized residual is a little unstable, and the Cook distance still can identify the outliers. As the best inference, the estimate based on LMS initial values can correctly identify outliers (Figure 8.8).

```
fittAll <- function(str) {
  list(
    lakes_ols =
      nlr(
        nlrobj5[[2]],
        data = list(
          xr1 = Lakes$nin,
          xr2 = Lakes$tw,
          yr = Lakes$tn
        ),
        start = str,
        control = nlr.control(
          method = "OLS",
          tolerance = 1e-6,
          maxiter = 2000
        )
      ),

    lakes_qt =
      nlrq(
        nlrobj5[[2]]$origin,
        data = list(
          xr1 = Lakes$nin,
          xr2 = Lakes$tw,
          yr = Lakes$tn
        ),
        start = str
      ),
    lakes_irls  =
      nlrob(
        nlrobj5[[2]]$origin,
        data = list(
          xr1 = Lakes$nin,
          xr2 = Lakes$tw,
          yr = Lakes$tn
        ),
        trace = TRUE,
        start = str,
        maxi = 5000
      ),

    lakes_nls  =
      nls(
        nlrobj5[[2]]$origin,
```

```
        data = list(
          xr1 = Lakes$nin,
          xr2 = Lakes$tw,
          yr = Lakes$tn
        ),
        trace = TRUE,
        start = str,
        control = nls.control(maxiter = 5000)
      ),

    lakes_MM =
      nlr(
        nlrobj5[[2]],
        data = list(
          xr1 = Lakes$nin,
          xr2 = Lakes$tw,
          yr = Lakes$tn
        ),
        control = nlr.control(
          tolerance = 1e-10,
          maxiter = 5000,
          initials = "lms"
        )
      ),
    lakes_MM2 =
      nlr(
        nlrobj5[[2]],
        data = list(
          xr1 = Lakes$nin,
          xr2 = Lakes$tw,
          yr = Lakes$tn
        ),
        start = str,
        control = nlr.control(tolerance = 1e-10, maxiter = 5000)
      )
    )
}
lakes_lms <-
  nlr(
    nlrobj5[[2]],
    data = list(
      xr1 = Lakes$nin,
      xr2 = Lakes$tw,
      yr = Lakes$tn
    ),
    start = list(p1 = 1, p2 = 1),
    control = nlr.control(
      method = "lms",
      tolerance = 1e-6,
      maxiter = 2000
    )
  )
```

```
library(robustbase)
str = lakes_lms$parameters
lkfit_lmas<-fittAll(str)
str = list(p1=1,p2=1)
lkfit_manual<-fittAll(str)
```

9.2   Simulated Data Examples

To assess the capabilities of the packages discussed in the last section, a simulation study is needed, so as to reveal their biases. Consider the artificially contaminated data in Figure 3.1. The data are fitted using six different methods using three functions of the three packages:

- quantile regression using the nlrq function from the quantreg package
- robust M-estimate using the nlrob function from the robustbase package
- OLS estimate using the nls function from the stats package
- other methods, such as OLS, LMS and MM-estimates, from the nlr package.

Table 9.2 reveals that the nlrq has smallest biase for p_1, p_3 and MM estimate by nlr has smallest bias for p_2. The OLS by nlr and nls functions

Table 9.2 Parameter estimates for artificially contaminated data, using nlr, nlrq, nlrob, and nls functions, and parameter biases.

Estimates	p_1	p_2	p_3	sigma
Exact	2575.1	41.3607	0.1118	210.074
nlr-lms	2296.4	38.0549	0.1166	–
nlr-OLS	1554.9	3445828	1.0066	329.579
nlr-MM	2515.8	45.7783	0.1161	43.165
nlrq	2587.8	33.0455	0.1055	–
nlrob-M	1591.1	770885	0.9055	208.556
nls-OLS	1554.9	3445355	1.0066	329.579

Biases	p_1	p_2	p_3	sigma
nlr-lms	−278.7	−3.3057	0.0047	–
nlr-OLS	−1020.1	3445786	0.8947	119.505
nlr-MM	−59.2	4.418	0.0043	−166.909
nlrq	12.7	−8.315	−0.0064	–
nlrob-M	−983.9	770844	0.7937	−1.518
nls-OLS	−1020.1	3445313	0.8947	119.505

Initial values are least median square, and manually fixed values (p_1=1, p_2=1)

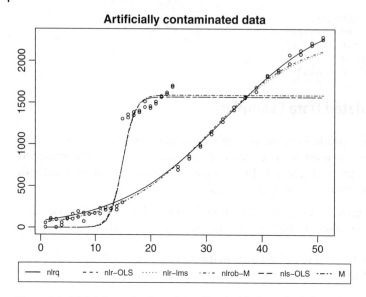

Artificially contaminated data

Figure 9.2 Artificially contaminated data fitted with six methods from three packages.

dramatically affected by outliers, meanwhile the M-estimate by `nlrob` function is unexpectedly broken down as well.

Figure 9.2 shows the fitted model by six methods and all three packages. The OLS estimate using `nlr` function and OLS by `nls` function are very close in the graph such that their difference can not be seen in the curves. The M-estimate using `nlrob` function is broken down as well. The predicted line by quantile estimates using `nlrq` fits the data better than all, and is close to LMS, MM, estimates using `nlr` function.

The commands below compute the six estimates and produce Figure 9.2.

```
library(robustbase)
library(quantreg)
library(nlr)
lodata <- list(xr = nlr::lotsout[, 1], yr = nlr::lotsout[, 2])
fitt.LO <- list(NULL)
fitt.LO[[1]] <-
  nlrq(
    nlrobj1[[14]]$origin,
    data = lodata,
    start = list(p1 = 2100, p2 = 38, p3 = .11),
    tau = 0.5,
    trace = TRUE
  )
fitt.LO[[2]] <-
  nlr(
    formula = nlrobj1[[14]],
```

```
      data = lodata,
      start = list(p1 = 2100, p2 = 38, p3 = .11),
      control = nlr.control(method = "OLS", maxiter = 200)
    )
fitt.LO[[3]] <-
  nlr(
      formula = nlrobj1[[14]],
      data = lodata,
      start = list(p1 = 2100, p2 = 38, p3 = .11),
      control = nlr.control(method = "lms", maxiter = 200)
    )
fitt.LO[[4]] <-
  nls(
      nlrobj1[[14]]$origin,
      data = lodata,
      start = list(p1 = 2100, p2 = 38, p3 = .11),
      control = nls.control(maxiter = 5000)
    )
fitt.LO[[5]] <-
  nlr(
      formula = nlrobj1[[14]],
      data = lodata,
      start = list(p1 = 2100, p2 = 38, p3 = .11),
      control = nlr.control(initials = "lms", maxiter = 200)
    )
fitt.LO[[6]] <-
  nlrob(
      nlrobj1[[14]]$origin,
      data = lodata,
      start = list(p1 = 2100, p2 = 38, p3 = .11),
      trace = T,
      control = nls.control(maxiter = 5000)
    )

xr = fitt.LO[[2]]$data$xr
yr = fitt.LO[[2]]$data$yr
yr1 <- predict(fitt.LO[[1]])  #nlrq
yr2 <- predict(fitt.LO[[2]])  #nlr-ols
yr3 <- predict(fitt.LO[[3]])  #nlr-lms
yr4 <- predict(fitt.LO[[4]])  #nls-ols
yr5 <- predict(fitt.LO[[5]])  #nlr-MM
yr6 <- predict(fitt.LO[[6]])  #nlrob-M

par(xpd = NA)
plot(xr,
     yr,
     main = "Artificially Contaminated Data",
     xlab = "",
     ylab = "")
lines(xr, yr1, lty = 1)
lines(xr, yr2, lty = 2)
```

```
lines(xr, yr3, lty = 3)
lines(xr, yr4, lty = 5)
lines(xr, yr5, lty = 6)
lines(xr, yr6, lty = 6)
legend(
  -1,-550,
  legend = c("nlrq", "nlr-OLS", "nlr-lms", "nls-OLS",
  "nlr-MM", "nlrob-M"),
  lty = 1:6,
  cex = .6,
  box.lwd = .7,
  lwd = 1,
  horiz = T,
  bty = "o"
)
```

A

nlr Database

This appendix presents the variables created as databases to store the different kinds of objects needed in nlr, and used in this book. There are two kind of databases supplied in nlr: sample data and lists of nonlinear functions of type nl.form objects. The latter include nonlinear regression models, robust loss functions, and nonlinear variance models.

The sample data are the real examples or simulated data used in the book. Mostly they follow nonlinear patterns, and will be used to fit appropriate nonlinear regression functions. Each data set may be fit with one or more nonlinear regression functions. Some of them include outliers, robust fits for which are discussed in the book.

Since nl.form is a general object that stores nonlinear functions, it is flexible enough to store different kind of algebraic functions. Several nonlinear function models are stored in nlr databases as lists of the nl.form object type. There are three kinds of nonlinear model stored in the nlr databases:

- nonlinear regression models, $f(x; \theta)$, that describe the data curve pattern
- loss functions ρ, used in minimizing robust loss object functions
- heteroscedastic variance nonlinear functions.

Although these are all stored in the nl.form object, in detail nlr uses them in slightly different ways, as explained in Chapter 8.

A.1 Data Set used in the Book

The data sets described in this section are embedded in the nlr package. They can be accessed by appending the name with a colon to nlr::, or by the data command followed by name of the variables, which all are exactly same as their titles. The data names that can be used are: carbon, methane, weights, Lakes, net.ir, net.kw, net.ch, net.sw, ntp.

Robust Nonlinear Regression: with Applications using R, First Edition.
Hossein Riazoshams, Habshah Midi, and Gebrenegus Ghilagaber.
© 2019 John Wiley & Sons Ltd. Published 2019 by John Wiley & Sons Ltd.
Companion website: www.wiley.com/go/riazoshams/robustnonlinearregression

A.1.1 Chicken-growth Data

The chicken-growth data were collected by Riazoshams and Miri (2005), for a period of 51 days of broiler chicken supply in an area of Marvdasht, Fars province, Iran. On the first day recorded, the population was 7300 chicken, housed in a growth chamber. The data were collected at random from this population. For the first 25 days, 10 chicken were selected randomly from the growth house and the weight of each was recorded. From the 27th day until the 51st (last day), 20 chicken were collected every two days, and the weight of each chicken was recorded, as shown in Table A.1.

At each date, the chickens were collected randomly from a different part of the chamber. The error of sampling can be measured for each date, but the most important date of production was the inflection point of the nonlinear curve, corresponding to the highest rate of chicken growth between the 14th and 20th days. For example, on the 16th day, the weight average was 306.5 g, with a sample variance equal to 1634.504 and standard deviation of 40.429 g. The variance of mean therefore equals $Var(\bar{x}) = 163.226$ with standard deviation 12.776. In conclusion, the relative precision of the mean will be 0.042, which means the error of sampling is 4.2%, which is acceptable. For the last day (day 51) the relative precision is 0.043. In conclusion, the error of the computations is around 4.5%, or a precision of 96.5%, which is adequate precision.

The data object supplied for the chicken-growth data in `nlr` package can be retrieved using the R command:

```
> data(Weights)
```

or direct use of variable name `nlr::Weight`. The object includes the predictor `Date` and response `Weight`, which can be accessed by R commands:

```
> Weights\$Date
> Weights\$Weight
```

The chicken growth depends on environmental, bioeconomic and economic factors, as well as the chicken strain and gender, but here we only derive the model for overall population weight. The model is nonlinear, and naturally, due to increases of weight over time, has heterogeneous variance. A logistic model with the variance as a power function of the model is fitted. The logistic nonlinear regression model can be accessed from the `nlrobj1[[14]]` variable of the `nl.form` object type. The power variance model can be accessed from the `nlrobjvarmdls1[[1]]` variable of the `nl.form` object type using the command `dt.3case[[3]]`.

A.1.2 Environmental Data

UNEP (1989) presented the data for methane and carbon dioxide gas concentrations obtained from gas trapped in ice at the South Pole 8000 years ago.

Table A.1 Chicken-growth data (g).

Date: 1	2	3	4	5	6	7	8	9	10	11	12	13	14	15	16	17	18	19	20	21
43	47	62	79	79	85	107	98	142	187	217	230	265	335	348	287	395	417	575	640	580
47	52	67	67	80	108	104	122	119	173	213	220	240	330	318	282	475	497	449	600	505
43	60	62	80	77	90	92	126	127	187	210	185	298	337	317	330	415	362	480	607	630
40	53	60	70	77	98	87	100	134	166	150	165	257	320	293	357	330	492	520	612	665
41	48	57	67	87	92	107	114	139	119	140	191	260	225	222	332	425	545	335	462	540
47	48	57	75	75	88	109	113	144	109	148	194	150	245	295	327	250	430	492	592	575
41	50	62	64	76	96	103	115	147	167	180	213	180	270	275	357	362	345	457	392	470
45	52	67	69	92	106	97	84	102	120	170	192	260	221	338	242	280	458	392	455	580
45	56	62	74	90	86	102	87	133	107	130	205	187	205	202	297	369	400	437	417	605
55	57	61	72	68	70	114	80	122	127	135	195	205	224	277	254	314	352	420	427	610

Date: 22	23	24	25		27		29		31		33		35		37		39		41
770	712	740	920	985	1045	945	1045	1215	885	1255	1430	1215	1070	1575	1300	2005	1685	1980	1875
685	770	675	840	950	725	965	1075	1125	1025	1325	1030	1378	1195	1800	1280	1980	1900	2020	2100
750	692	895	777	860	735	1075	1020	1190	1110	985	1065	1725	1400	1510	1175	1955	1785	1705	1795
717	692	775	860	875	675	1070	1155	1175	1125	1335	1260	1475	1575	1500	1250	1380	2000	1430	2000
557	530	720	640	775	735	885	1230	925	1115	1345	1260	1475	1600	1330	1800	2000	1825	1370	1925
402	660	550	530	865	805	890	800	1310	1240	1425	1465	1800	1380	1430	1905	1130	1605	2050	1480
470	585	620	775	665	835	970	895	1270	1165	930	1545	1205	1480	1600	1480	1150	1920	1785	2100
490	670	870	790	750	690	1025	965	1005	1110	1070	910	1300	1720	1830	1725	1950	1270	2000	2080
390	660	825	875	1065	785	940	845	1035	850	1185	925	1005	1585	1600	1375	1685	1880	1250	2055
425	800	800	840	970	806	1070	1110	1150	1115	1180	935	975	1320	1720	1480	1900	1330	1230	2105

Date: 43		45		47		49		51											
1575	2250	1630	2050	1430	2060	2220	2235	2055	1420	2100	1895	2100	2145	2095	2145	2140	1545	2200	2205
2090	1268	2160	2245	1930	1770	2240	2205	2610	2110	2440	2300	2060	2095	1515	2760	2705	2145	2535	2395
1725	1850	2170	2235	1830	2220	1385	2450	2735	2120	2870	1745	2655	2490	2075	2320	2850	2150	3005	1830
2135	2175	1685	1595	1860	1975	2010	2230	1500	1755	2840	1800	1795	2900	1825	1835	1585	1875	2940	1910
2225	2160	2355	1690	2000	2170	1765	2225	1780	2015	2030	2710	2780	2915	1455	1805	1895	2120	2130	2840

Source: Riazoshams and Miri (2005). Reproduced with permission of Islamic Azad University, Abadeh.

The methane gas data are shown in Table A.2, and the carbon dioxide data are shown in Table A.3).

The data object supplied for the methane data in the `nlr` package can be retrieved using the R command:

```
> data(methane)
```

or direct use of variable name `nlr::methane`. The object includes the predictor `year` and the response `methane`, which can be accessed by R commands:

```
> methane$year
> methane$methane
```

The data object supplied for the carbon dioxide data in the `nlr` package can be retrieved using the R command:

```
> data(carbon)
```

or by direct use of variable name `nlr::carbon`. The object includes the predictor `year` and the response `co2`, which can be accessed by R commands:

```
> carbon$year
> carbon$co2
```

Four exponential forms were fitted to the methane and carbon dioxide data. The models can be accessed from the `nlrobj5[[4]]`, `nlrobj5[[8]]`, `nlrobj5[[9]]` and `nlrobj5[[18]]` variables of the `nl.form` object type.

A.1.3 Lakes Data

Table A.4 shows the lakes data set, which was collected from 29 lakes in Florida by the United States Environmental Protection Agency (1978). The data presents the relationship between mean annual total nitrogen concentration (TN), as the response variable, and the average influence nitrogen concentration (NIN) and water retention time (TW), as predictors. Stromberg (1993) has identified observations 10 and 23 as outliers.

The data object supplied for the lakes data in the `nlr` package can be retrieved using the R command:

```
> data(Lakes)
```

or by direct use of the variable name `nlr::Lakes`. The object includes predictors `nin`, `tw`, and response `tn`, which can be accessed by R commands:

```
> Lakes$nin
> Lakes$tw
>Lakes$tn
```

The two variate fraction function is ftited to the lakes data, can be accessed from the `nlrobj5[[2]]` variables of the `nl.form` object type.

Table A.2 Methane data.

year	methane	year	methane	year	methane	year	methane
−7788	760	1234	670	1774	830	1940	1110
−1478	710	1284	690	1774	770	1949	1160
−1066	590	1294	920	1784	850	1950	1180
−926	520	1344	640	1794	840	1954	1260
−826	630	1374	890	1794	820	1955	1210
−306	570	1374	680	1794	780	1955	1400
24	650	1404	760	1804	730	1955	1300
114	630	1414	800	1804	840	1956	1310
154	820	1444	690	1814	770	1956	1340
178	650	1444	660	1819	980	1965	1388
194	620	1474	630	1827	830	1966	1340
194	640	1474	730	1829	730	1967	1480
254	640	1514	660	1834	860	1968	1370
354	640	1524	720	1839	930	1969	1380
424	680	1564	650	1844	790	1970	1335
444	660	1564	600	1844	780	1971	1400
524	640	1564	700	1849	890	1972	1448
534	600	1574	750	1857	900	1973	1554
664	670	1600	860	1861	830	1974	1480
694	680	1624	670	1869	870	1975	1450
694	620	1624	720	1874	890	1975	1470
734	650	1654	640	1880	950	1976	1500
764	800	1654	650	1882	900	1976	1488
774	630	1664	800	1884	1070	1977	1480
794	650	1664	750	1884	880	1977	1509
794	620	1699	740	1884	920	1978	1555
864	690	1704	730	1893	870	1978	1530
874	700	1714	680	1904	1130	1979	1565
894	650	1724	710	1904	970	1979	1551
924	690	1724	750	1907	990	1980	1590
964	770	1739	690	1910	1000	1980	1569
964	700	1754	790	1917	990	1981	1580
974	690	1764	770	1918	1090	1981	1587
1054	680	1769	710	1919	1000	1982	1601
1144	700	1771	780	1925	1020	1983	1611
1204	630	1771	800	1927	1100	1984	1626

Source: UNEP 1989 Environmental data report prepared for UNEP by the GEMS Monitoring and Assessment, Research Centre. London, UK, in co-operation with the World Resources Institute, Washington, D.C. Adapted from UNEP 1989.

Table A.3 Carbon dioxide data.

year	CO_2	year	CO_2	year	CO_2	year	CO_2
1754	279	1839	283.1	1943	307.9	1974	330.41
1772	279	1843	287.4	1953	312.7	1975	331.01
1806	280	1847	286.8	1959	316.1	1976	332.06
1825	284	1854	288.2	1960	317.02	1977	333.62
1864	288	1869	289.3	1961	317.74	1978	335.19
1894	297	1874	289.5	1962	318.63	1979	336.54
1914	300	1878	290.3	1963	319.13	1980	338.4
1932	306	1887	292.3	1965	320.41	1981	339.46
1949	311	1899	295.8	1966	321.09	1982	340.76
1958	312	1903	294.8	1967	321.9	1983	342.76
1965	318	1905	296.9	1968	322.72	1984	344.34
1973	328	1909	299.2	1969	324.21	1985	345.65
1744	276.8	1915	300.5	1970	325.51	1986	346.84
1764	276.7	1921	301.6	1971	326.48	1987	348.62
1791	279.7	1927	305.5	1972	327.6		
1816	283.8	1935	306.6	1973	329.82		

Source: UNEP 1989 Environmental data report prepared for UNEP by the GEMS Monitoring and Assessment, Research Centre. London, UK, in co-operation with the World Resources Institute, Washington, D.C. Adapted from UNEP 1989.

Table A.4 Lakes data.

nin:	tn:	tw:	nin:	tn:	tw:
5.548	2.59	0.137	5.011	4.77	0.099
4.896	3.77	2.499	2.455	2.22	0.644
1.964	1.27	0.419	0.913	0.59	0.266
3.586	1.445	1.699	0.89	0.53	0.351
3.824	3.29	0.605	2.468	1.91	0.027
3.111	0.93	0.677	4.168	4.01	0.03
3.607	1.6	0.159	4.81	1.745	3.4
3.557	1.25	1.699	34.319	1.965	1.499
2.989	3.45	0.34	1.531	2.555	0.351
18.053	1.096	2.899	1.481	0.77	0.082
3.773	1.745	0.082	2.239	0.72	0.518
1.253	1.06	0.425	4.204	1.73	0.471
2.094	0.89	0.444	3.463	2.86	0.036
2.726	2.755	0.225	1.727	0.76	0.721
1.758	1.515	0.241			

nin: average influent nitrogen concentration, tn: mean annual total nitrogen concentration, tw: water retention time.
Source: Reproduced with permission of Taylor & Francis.

A.1.4 Economic Data

Economic data in `nlr` include the Net Money and Trademark data sets. The net money data are shown in Table A.5, and were collected from http://www .worldbank.com/, for four countries – China, Iran, Kuwait and Sweden – from 1960 to 2012. The China, Iran and Kuwait data can be expressed as a nonlinear power function model, or a transformed logarithm linear model. In addition, the Iran and Kuwait data include outliers.

The data object supplied for the net money data in the `nlr` package can be retrieved using the R commands:

```
> data(net.ir) # Iran Net Money Data
> data(net.ch) # China Net Money Data
> data(net.kw) # Kuwait Net Money Data
> data(net.sw) # Sweeden Net Money Data
```

or by direct use of variable names `nlr::net.ir`, `nlr::net.ch`, `nlr::net.kw`, `nlr::net.ch`. The object includes the predictor `year` and the response `net`, for all variables, which can be accessed by R commands:

```
> net.ir$year
> net.ir$net
> net.ch$year
> net.ch$net
> net.kw$year
> net.kw$net
> net.sw$year
> net.sw$net
```

Table A.6 display Iran trademark applications (direct resident) data. It includes outliers at the 44th point (2003). The errors of a nonlinear regression model fitted to these data is autocorrelated.

The data object supplied for the Iran trademark data in the `nlr` package can be retrieved using the R command:

```
> data(trade.ir)
```

or by direct use of variable name `nlr::trade.ir`. The object includes the predictor `year` and the response `trade`, for all variables, which can be accessed by R commands:

```
> trade.ir$year
>trade.ir$trade
```

Four exponential forms fitted to the net money and trademark data sets, can be accessed from the `nlrobj5[[4]]`, `nlrobj5[[8]]`, `nlrobj5[[9]]`, `nlrobj5[[18]]` variables of the `nl.form` object type.

Table A.5 Net money data.

Year	net.chd	net.ird	net.kwd	net.swd	Year	net.chd	net.ird	net.kwd	net.swd
1960	*	6.12E+10	3.80E+06	4.01E+10	1987	9.71E+11	1.34E+13	4.72E+09	6.65E+11
1961	*	5.92E+10	5.00E+05	4.16E+10	1988	1.15E+12	1.58E+13	5.39E+09	7.97E+11
1962	*	6.88E+10	1.48E+07	4.46E+10	1989	1.35E+12	1.92E+13	5.59E+09	9.43E+11
1963	*	8.06E+10	3.24E+07	4.92E+10	1990	1.67E+12	2.28E+13	*	9.51E+11
1964	*	1.01E+11	2.29E+07	5.27E+10	1991	2.00E+12	2.80E+13	5.78E+09	9.41E+11
1965	*	1.13E+11	2.74E+07	5.77E+10	1992	2.45E+12	3.39E+13	6.21E+09	8.55E+11
1966	*	1.31E+11	3.66E+07	6.27E+10	1993	3.58E+12	4.34E+13	6.05E+09	8.72E+11
1967	*	1.57E+11	4.80E+07	7.08E+10	1994	4.31E+12	5.95E+13	6.61E+09	8.72E+11
1968	*	1.91E+11	7.69E+07	7.95E+10	1995	5.33E+12	7.44E+13	7.21E+09	8.35E+11
1969	*	2.36E+11	7.37E+07	8.53E+10	1996	6.64E+12	1.02E+14	7.15E+09	8.33E+11
1970	*	3.04E+11	4.15E+07	8.92E+10	1997	7.95E+12	1.16E+14	8.08E+09	9.09E+11
1971	*	3.48E+11	3.29E+07	9.74E+10	1998	9.55E+12	1.52E+14	8.46E+09	9.98E+11
1972	*	4.27E+11	6.80E+07	1.10E+11	1999	1.07E+13	1.87E+14	8.48E+09	9.73E+11
1973	*	5.94E+11	1.61E+08	1.22E+11	2000	1.19E+13	2.11E+14	8.15E+09	1.10E+12
1974	*	5.20E+11	5.49E+07	1.39E+11	2001	1.35E+13	2.47E+14	8.70E+09	2.50E+12
1975	*	8.38E+11	1.42E+08	1.58E+11	2002	1.73E+13	3.20E+14	9.48E+09	2.60E+12
1976	*	1.37E+12	5.80E+08	1.70E+11	2003	2.06E+13	4.17E+14	1.09E+10	2.70E+12
1977	1.26E+11	1.66E+12	7.82E+08	1.89E+11	2004	2.24E+13	5.14E+14	1.14E+10	2.88E+12
1978	1.39E+11	*	1.17E+09	2.27E+11	2005	2.48E+13	6.01E+14	1.31E+10	3.25E+12
1979	1.98E+11	2.72E+12	1.50E+09	2.69E+11	2006	2.89E+13	7.19E+14	1.77E+10	3.60E+12
1980	2.42E+11	4.24E+12	1.80E+09	3.07E+11	2007	3.40E+13	1.02E+15	2.27E+10	4.08E+12
1981	2.74E+11	5.69E+12	2.68E+09	3.64E+11	2008	3.79E+13	1.06E+15	2.62E+10	4.31E+12
1982	3.05E+11	7.13E+12	3.18E+09	4.02E+11	2009	4.95E+13	1.22E+15	2.76E+10	4.48E+12
1983	3.44E+11	7.14E+12	3.22E+09	4.36E+11	2010	5.87E+13	7.81E+14	2.49E+10	4.74E+12
1984	4.51E+11	*	3.33E+09	5.01E+11	2011	6.88E+13	*	2.52E+10	4.98E+12
1985	5.93E+11	*	3.32E+09	5.05E+11	2012	8.06E+13	*	2.50E+10	5.16E+12
1986	7.94E+11	1.09E+13	3.85E+09	5.83E+11					

*missing value, net.chd, China net money; net.ird, Iran net money data; net.kwd, Kuwait net money data; net.swd, Sweden net money.

Table A.6 Iran trademark data.

year	trademark	year	trade	year	trade	year	trade
1960	895	1972	1967	1984	1542	1996	5021
1961	905	1973	1733	1985	1681	1997	6145
1962	786	1974	1542	1986	1624	1998	5417
1963	1145	1975	1374	1987	1651	1999	8723
1964	1246	1976	1787	1988	1439	2000	9173
1965	1608	1977	1336	1989	1708	2001	9858
1966	1685	1978	1553	1990	2195	2002	11784
1967	1369	1979	769	1991	2533	2003	7468
1968	1670	1980	1186	1992	2929	2004	17607
1969	1806	1981	1987	1993	4573	2005	20439
1970	1748	1982	2290	1994	5936	2006	23827
1971	1681	1983	2238	1995	6184		

Iran trademark applications, direct residents.
Source: Reproduced with permission of World Bank.

A.1.5 National Texicology Program (NTP) Data

Table A.7 shows the mouse kidney data from National Toxicology Study Program (Bucher (2007)). The data object supplied in the `nlr` package can be retrieved using the R command:

```
> data(ntp)
```

or by direct use of variable name `nlr::ntp`. The object includes the dose concentration predictor `dm.k`, and the chromium concentration response `cm.k`, which can be accessed by R commands:

```
> ntp$cm.k
> ntp$dm.k
```

The Hill model fitted to data by Lim et al. (2010, 2012), can be accessed from the `nlrobj1[[16]]` variable of the `nl.form` object type. The fraction of exponential form for the variance model can be accessed from the `nlrobj-varmdls3[[2]]` variable of the `nl.form` object type.

A.1.6 Cow Milk Data

Table A.8 displays the milk production of a single cow, as measured in Iran. Silvestre et al. (2006) have fitted five models to this data set: the Wood, Wilmink, Ali and Scheffer, cubic splines, and Legendre polynomials models.

Table A.7 ntp data.

Dose	Chromium	Dose	Chromium
0	0.072	10	0.67
0	0.114	10	0.583
0	0.113	30	1.2
0	0.121	30	1.41
1	0.243	30	1.66
1	0.168	30	1.3
1	0.182	100	3.93
1	0.22	100	5.31
3	0.339	100	5.68
3	0.374	100	2.29
3	0.339	300	2.29
3	0.346	300	2.67
10	0.673	300	4.42
10	0.797	300	3.64

Predictor: dose, response: chromium concentration in mouse kidney.
Source: Reproduced with permission of Springer.

Table A.8 Cow milk production data for a single cow in a year.

Day	Milk
15	25.6
49	40.2
77	42.0
110	43.8
140	42.0
170	43.6
197	41.8
226	43.8
255	42.0
285	42.0

Predictor: day, response: milk produced (l).
Source: Reproduced with permission of Emaane Nooraee.

The data object supplied for the cow milk data in the `nlr` package can be retrieved using the R command:

```
> data(cow)
```

or by direct use of the variable name `nlr::cow`. The object includes the day predictor `Day`, and the milk response `Milk`, which can be accessed by R commands:

```
> cow$Day
> cow$Milk
```

Two nonlinear regression models, the Wood model and the Wilmink model, are provided in the `nlr` package, stored as the `nlrobj[[9]]` and `nlrobj[[10]]` variables in the `nl.form` object, respectively.

A.1.7 Simulated Outliers

The simulated outliers data used in Chapter 6 for assessing statistical measure performance is shown in Table A.9. The data are simulated from the logistic model

$$y_i = \frac{a}{1 + b.e^{-cx_i}} + \varepsilon_i$$

where $\theta = (a = 2575, b = 41, c = 0.11)$, $\varepsilon_i \sim N(0, \sigma^2 = 70^2)$, and $x_i \sim U[3, 50]$. Three cases of outliers are artificially created.

- Case A. The first good datum point (x_1, y_1), is replaced with a new value $y_1 + 1000$.
- Case B. The 6th, 7th and 8th data points are replaced, with their corresponding y values increased by 1000 units.
- Case C. Six high leverage points were created by replacing the last six observations with (x, y) pair values of $(90, 6500)$, $(92, 6510)$, $(93, 6400)$, $(93, 6520)$, $(90, 6600)$, $(94, 6600)$.

The data object supplied for the simulated data in the `nlr` package can be retrieved using the R command:

```
> data(dt.3case)
```

or by direct use of variable name `nlr::dt.3case[[3]]`. The object includes an array of lists of several simulated cases. The three cases, stored in index 3, 4, 10, for Cases A, B, and C respectively, can be retrieved by the R commands:

```
>dt.3case[[3]]
>dt.3case[[4]]
>dt.3case[[10]]
```

Table A.9 Three cases of simulated outliers from logistic model.

Index	x_i	Case A y_i	Case B y_i	Index	Case C x_i	y_i
1	3.0000	1084.1991	84.1991	1	3.000	84.199
2	5.4737	247.4617	247.4617	2	5.474	247.462
3	7.9474	17.6354	17.6354	3	7.947	17.635
4	10.4211	114.7414	114.7414	4	10.421	114.741
5	12.8947	323.0059	323.0059	5	12.895	323.006
6	15.3684	363.4110	1363.4110	6	15.368	363.411
7	17.8421	327.9053	1327.9053	7	17.842	327.905
8	20.3158	373.8674	1373.8674	8	20.316	373.867
9	22.7895	483.5325	483.5325	9	22.789	483.532
10	25.2632	713.5400	713.5400	10	25.263	713.540
11	27.7368	836.5337	836.5337	11	27.737	836.534
12	30.2105	1041.9627	1041.9627	12	30.211	1041.963
13	32.6842	1286.4226	1286.4226	13	32.684	1286.423
14	35.1579	1363.4546	1363.4546	14	35.158	1363.455
15	37.6316	1629.1149	1629.1149	15	37.632	1629.115
16	40.1053	1813.2731	1813.2731	16	40.105	1813.273
17	42.5789	1817.2337	1817.2337	17	42.579	1817.234
18	45.0526	1975.5210	1975.5210	18	45.053	1975.521
19	47.5263	2002.3536	2002.3536	19	47.526	2002.354
20	50.0000	2243.2583	2243.2583	20	50.000	2243.258
				21	90.000	6500.000
				22	92.000	6510.000
				23	93.000	6400.000
				24	93.000	6520.000
				25	90.000	6600.000
				26	94.000	6600.000

Predictor values for Case A and B are the same.

The predictor variable xr, and response yr, be accessed by R commands:

```
> dt.3case[[3]]$xr
> dt.3case[[3]]$yr
> dt.3case[[4]]$xr
> dt.3case[[4]]$yr
> dt.3case[[10]]$xr
> dt.3case[[10]]$yr
```

Other indexes include simulation cases that are not discussed in the book, but used for testing purposes by `nlr`. The logistic nonlinear regression model can be accessed from the `nlrobj1[[14]]` variable of the `nl.form` object type. The power variance model can be accessed from the `nlrob-jvarmdls1[[1]]` variable of the `nl.form` object using the command `type.dt.3case[[3]]`.

A.1.8 Artificially Contaminated Data

The artificially contaminated data used in Chapter 3 and 9 were created for research and comparison purposes. The data are shown in Table A.10. Several outliers in the middle are created by shifting values up (see Figures 3.1 and 9.2).

The data are simulated from the logistic model

$$y_i = \frac{a}{1 + b.e^{-cx_i}} + \varepsilon_i$$

where $\theta = (a = 2575, b = 41, c = 0.11)$, $\varepsilon_i \sim N(0, \sigma^2 = 70^2)$, and $x_i \sim U[3, 50]$. These values are selected to mimic the real chicken-growth data, as shown in Table A.1. The parameter values are selected from the OLS estimates of the data without considering heteroscedasticity.

The data object for the simulated data in `nlr` package can be retrieved using the R command:

```
> data(lotsout)
```

or directly by use of variable name `nlr::lotsout`. In order to mimic real life, the scale of the dependent variables is considered equal to the weight data. They are repeated twice and stored in the first column of `lotsout` variable. The response values are simulated and stored in second column of the `lotsout` variable.

In the middle of the data, several values are raised by a large fixed value to create artificial outlier points (See Figures 3.1 and 9.2).

A.2 Nonlinear Regression Models

A set of six variables databases, each including several nonlinear regression function models, is supplied in `nlr`. These are called `nlrobj1`, `nlrobj3`, `nlrobj4`, `nlrobj5`, `nlrobj6` and `nlrobj7`. Some of them were constructed by Bunke et al. (1995b, 1998) and converted to `nl.form` objects by adding gradient, Hessian and other slots to the new object. The data objects supplied for the simulated data in the `nlr` package can be retrieved using the R commands:

Table A.10 Artificially contaminated data.

x	y	x	y	x	y	x	y
1	60.99	10	178.56	20	1462.31	33	1311.23
1	9.98	11	237.73	20	1439.84	35	1418.36
2	98.75	11	170.83	21	1493.14	35	1455.05
2	115.61	12	246.5	21	1517.5	37	1563.39
3	7.51	12	217.33	22	1573.55	37	1571.21
3	105.46	13	229.27	22	1584.56	39	1638.21
4	65.27	13	233.03	23	1608.05	39	1689.63
4	30.82	14	218.83	23	1627.7	41	1833.68
5	107.33	14	259.34	24	1711.54	41	1819.2
5	119.35	15	305.59	24	1694.8	43	1895.1
6	167	15	1310.42	25	700.05	43	1875.16
6	106.89	16	1324.71	25	737.38	45	2079.48
7	130.22	16	1362.85	27	861.67	45	1975.91
7	198.74	17	1390.56	27	834.79	47	2131.97
8	181.04	17	1353.42	29	978.04	47	2087.82
8	79.73	18	1389.66	29	995.62	49	2223.17
9	164.9	18	1407.96	31	1158.54	49	2194.35
9	166.83	19	1519.65	31	1125.98	51	2294.72
10	176.79	19	1452.25	33	1274.69	51	2265.83

```
> data(nlrobj1)
> data(nlrobj3)
> data(nlrobj4)
> data(nlrobj6)
> data(nlrobj7)
```

or by direct use of variable name nlr::nlrobj1, or other indices. The objects include an array of lists of several nl.form nonlinear regression models. These are, respectively, objects 16, 18, 11, 12, and 23 in nlrobj1, nlrobj3, nlrobj4, nlrobj6 and nlrobj7. The objects stored in variables can be retrieved as arguments using R commands, for example:

```
>nlrobj1[[14]]
```

This command, retrieve nl.form object, contains the logistic nonlinear regression function model. Following Bunke et al. (1995b), the predictor variable is denoted xr, and the response is denoted yr.

A.3 Robust Loss Functions Data Bases

The robust ρ functions in the `nlr` package are stored in a list of `nl.form` called the `nl.robfuncs` variable. They can be accessed using the name of the variable and index brackets or can be read from the package into a variable. For example, `nl.robfuncs[[1]]` returns the *"hampel"* function. Table A.11 displays the robust ρ function, robust psi function (derivative of ρ), $\psi(t) = \rho'(t)$, the second derivative of the rho function $\rho''(t) = \psi'(t)$, default values for tuning parameters of each function, and the tuning constants k_0 and k_1 applied for calculating MM-estimates, as discussed in Chapter 3.11 and Example 8.3. The *"nl.robfuncs"* variables includes the weight of the rho function defined as $w(t) = \psi(t)/t$, for each of the loss functions.

Notes:

- At Index 5: *the "half huber"* function is equal to the *"huber"* function (index 1) divided by a constant number 2. All the functions in the other columns are divided by 2 and are not shown in the table. This is designed for special cases in which the authors use half of the function in computations.
- At Index 6 and 2: Index 6 includes the written formula and index 2 is a extended algebric form. The result is the same, but computation 2 might be slightly more efficient.
- At Index 7: *"least square"* is quadratic function that can be used for least squares estimation.

A.4 Heterogeneous Variance Models

A set of three variables databases, each including several nonlinear variance function models, is supplied in `nlr`. These are called `nlrobjvarmdls1`, `nlrobjvarmdls2`, and `nlrobjvarmdls3`. There are small differences: `nlrobjvarmdls1` returns variance of form $\sigma^2 g(x_i; \lambda)$ and will be used in subroutines that require variance in most cases. `nlrobjvarmdls2` returns standard deviation of form $\sigma\sqrt{g}$, and will be used in subroutines that require standard errors; it is used for compatibility, and the user must refer to the program documentation. `nlrobjvarmdls3` returns the general variance model $H(x_i; \tau)$.

The variance object supplied in the `nlr` package can be retrieved using the R commands:

```
> data(nlrobjvarmdls1)
> data(nlrobjvarmdls2)
> data(nlrobjvarmdls3)
```

Table A.11 Robust rho functions.

Index: function name	ρ function	ψ function	ρ''	Default values for Tuning constants
1: huber	$\rho(t) = \begin{cases} t^2 & \lvert t\rvert \le \alpha ; \\ 2\alpha\lvert t\rvert - \alpha^2 & \alpha < \lvert t\rvert . \end{cases}$	$\psi(t) = \begin{cases} 2t & \lvert t\rvert \le \alpha ; \\ 2\alpha\,\mathrm{sgn}(t) & \alpha < \lvert t\rvert . \end{cases}$	$\rho''(t) = \begin{cases} 2 & \lvert t\rvert \le \alpha ; \\ 0 & \alpha < \lvert t\rvert . \end{cases}$	$\alpha = 1.345$ $k_0 = 3.73677$ $k_1 = 4$ $\max \rho_0 = 1.345$
2 and 6***: hampel	$\rho(t) = \begin{cases} \dfrac{t^2}{2} & \lvert t\rvert \le \alpha ; \\[4pt] \alpha\left(\lvert t\rvert - \dfrac{\alpha}{2}\right) & \alpha < \lvert t\rvert \le \beta ; \\[6pt] \dfrac{\alpha\left(\gamma\lvert t\rvert - \dfrac{t^2}{2}\right)}{\gamma - \beta} - 7\dfrac{\alpha^2}{6} & \beta < \lvert t\rvert \le \gamma ; \\[8pt] \alpha(\beta + \gamma - \alpha)/2 & \gamma < \lvert t\rvert . \end{cases}$	$\psi(t) = \begin{cases} t & \lvert t\rvert \le \alpha ; \\[2pt] \alpha\,\mathrm{sgn}(t) & \alpha < \lvert t\rvert \le \beta ; \\[4pt] \dfrac{\alpha\,\mathrm{sgn}(t)}{\gamma - \beta}(\gamma - \lvert t\rvert) & \beta < \lvert t\rvert \le \gamma ; \\[4pt] 0 & \gamma < \lvert t\rvert . \end{cases}$	$\rho''(t) = \begin{cases} 1 & \lvert t\rvert \le \alpha ; \\ 0 & \alpha < \lvert t\rvert \le \beta ; \\ \dfrac{-\alpha}{\gamma - \beta} & \beta < \lvert t\rvert \le \gamma ; \\ 0 & \gamma < \lvert t\rvert . \end{cases}$	$\alpha = 1.5$ $\beta = 3.5$ $\gamma = 8$ $k_0 = 0.212$ $k_1 = 0.9014$ $\max \rho_0 = 3.75$
3: bisquare *	$\rho(t) = \begin{cases} 1 - \left(1 - \left(\dfrac{t}{\alpha}\right)^2\right)^3 & \lvert t\rvert \le \alpha ; \\ 1 & \alpha < \lvert t\rvert . \end{cases}$	$\psi(t) = \begin{cases} \dfrac{6}{\alpha^2}\left(t\left(1 - \left(\dfrac{t}{\alpha}\right)^2\right)^2\right) & \lvert t\rvert \le \alpha ; \\ 0 & \alpha < \lvert t\rvert . \end{cases}$	$\rho''(t) = \begin{cases} \dfrac{6}{\alpha^2} - 36\dfrac{t^2}{\alpha^4} + 30\dfrac{t^4}{\alpha^6} & \lvert t\rvert \le \alpha ; \\ 0 & \alpha < \lvert t\rvert . \end{cases}$	$\alpha = 4.685$ $k_0 = 1.56$ $k_1 = 4.68$ $\max \rho_0 = 1$
4: andrew	$\rho(t) = \begin{cases} \alpha\left(\alpha - \cos\left(\dfrac{t}{\alpha}\right)\right) & \lvert t\rvert \le \alpha\pi ; \\ \dfrac{\alpha^2}{2\alpha} & \alpha\pi < \lvert t\rvert . \end{cases}$	$\psi(t) = \begin{cases} \sin\left(\dfrac{t}{\alpha}\right) & \lvert t\rvert \le \alpha\pi ; \\ 0 & \alpha\pi < \lvert t\rvert . \end{cases}$	$\rho''(t) = \begin{cases} \cos\left(\dfrac{t}{\alpha}\right)/\alpha & \lvert t\rvert \le \alpha\pi ; \\ 0 & \alpha\pi < \lvert t\rvert . \end{cases}$	$\alpha = 1.339$
5: halph huber	"*halph huber*" is huber function devided by 2			
7: least square **	$\rho(t) = t^2$	$\psi(t) = 2\,t$	$\rho''(t) = 2$	

"`robfuncs`" variable includes "`nl.form`" object format of rho functions as presented in the index. (*) bisquare means "*bisquare*" function. (**) "least square" is a quadratic function that can be used for least square estimation. (***) index 6 is as written, index 2 is an algebrically extended formula.

Table A.12 Variance model functions.

Index: function name	H function	Index: function name	H function
1: Power	$H(t) = \sigma^2 t^\lambda$	2: Exponential	$H(t) = \sigma^2 \exp$
3: Linear	$H(t) = \sigma^2 * (1 + \lambda t)$	4: Unimodal quadratic	$H(t) = \sigma^2 + \lambda_1$ $(t - mt) + \lambda_2(t - mt)^2$
5: Bell shaped	$H(t) = \sigma^2 + \sigma^2$ $(max(t) - t)(min(t) - t)$	6: Simple linear	$H(t) = \sigma^2 + \lambda * (t - mt)$
7: Power no constant	$H(t) = t^\lambda$		

"nlrobjvarmdls1" variable includes nl.form object format of heteroscedastic variance model functions.

or by direct use of the variable name nlrobjvarmdls1, or other indices. The objects include an array of lists of several nl.form nonlinear regression models. There are, 7, 7, and 3 objects in nlrobjvarmdls3, nlrobjvar-mdls2, nlrobjvarmdls3, respectively. The objects stored in the variables can be retrieved as arguments by R commands, for example:

```
>nlrobjvarmdls1[[1]]
```

This command, retrieve nl.form object form, contains the heteroscedastic power nonlinear function model. Table A.12 shows the models stored in nlrobjvarmdls1.

References

Anscombe FJ and Tukey JW 1963 The examination and analysis of residuals. *Technometrics* **5** (2), 141–160.

Atkinson AC 1981 Two graphical displays for outlying and influential observations in regression. *Biometrika* **68** (1), 13.

Atkinson AC 1982 Regression diagnostics, transformations and constructed variables. *Journal of the Royal Statistical Society. Series B (Methodological)* **44** (1), 1–36.

Atkinson AC 1986 Masking unmasked. *Biometrika* **83** (3), 533.

Attar E, Vidyasagar RA and Dutta SRK 1979 An algorithm for l1-norm minimization with application to nonlinear l1-approximation. *SIAM Journal on Numerical Analysis* **16** (1), 70–86.

Barrodale I and Roberts F 1974 Solution of an overdetermined system of equations in the l 1 norm f4. *Communications of the ACM* **17** (6), 319–320.

Bartels RH and Conn AR 1980 Linearly constrained discrete l1 problems. *ACM Transactions on Mathematical Software (TOMS)* **6** (4), 594–608.

Bates DM and Watts DG 1980 Relative curvature measures of nonlinearity. *Journal of the Royal Statistical Society. Series B (Methodological)* **42** (1), 1–25.

Bates DM and Watts DG 2007 *Nonlinear Regression Analysis and its Applications.* Wiley Series in Probability and Mathematical Statistics: Applied Probability and Statistics. John Wiley and Sons Inc., New York.

Belsley DA, Kuh E and Welsch RE 1980 *Regression diagnostics: identifying influential data and sources of collinearity.* Wiley Series in Probability and Mathematical Statistics. John Wiley and Sons Inc., New York.

Bucher JR 2007 NTP technical report on the toxicity studies of sodium dichromate dihydrate (cas no. 7789-12-0) administered in drinking water to male and female F344/N rats and B6C3F1 mice and male BALB/c and am3-C57BL/6 mice. Technical report, National Toxicology Program, Research Triangle Park, North Carolina.

Bunke O, Droge B and Polzehl J 1995a Model selection, transformations and variance estimation in nonlinear regression.

Robust Nonlinear Regression: with Applications using R, First Edition.
Hossein Riazoshams, Habshah Midi, and Gebrenegus Ghilagaber.
© 2019 John Wiley & Sons Ltd. Published 2019 by John Wiley & Sons Ltd.
Companion website: www.wiley.com/go/riazoshams/robustnonlinearregression

Bunke O, Droge B and Polzehl J 1995b Splus tools for model selection in nonlinear regression. Discussion paper 95-73, Sonderforschungsbereich 373, Humboldt University.

Bunke O, Droge B and Polzehl J 1998 Splus tools for model selection in nonlinear regression. *Computational Statistics* **13** (2), 257–281.

Bunke O, Droge B and Polzehl J 1999 Model selection, transformations and variance estimation in nonlinear regression. *Statistics* **33** (3), 197–240.

Carroll RJ and Ruppert D 1988 *Transformation and weighting in regression*, vol. 30. CRC Press.

Charnes A, Cooper WW and Ferguson RO 1955 Optimal estimation of executive compensation by linear programming. *Management Science* **1** (2), 138–151.

Chen Y, Stromberg AJ and Zhou M 1997 The least trimmed squares estimate in nonlinear regression. Technical report. University of Kentucky.

Chong EK and Zak SH 1996 *An Introduction to Optimization*. John Wiley and Sons Inc., New York.

Čížek P 2001 Nonlinear least trimmed squares. Technical report, SFB Discussion Paper, Humboldt University, 25.

Cook RD 1986 Assessment of local influence. *Journal of the Royal Statistical Society. Series B (Methodological)* **48** (2), 133–169.

Cook RD and Weisberg S 1982 *Residuals and influence in regression*. Monographs on Statistics and Applied Probability, Vol. 18. Chapman and Hall, New York.

Cook RD, Tsai CL and Wei BC 1986 Bias in nonlinear regression. *Biometrika* **73** (3), 615.

DasGupta A 2008 *Asymptotic Theory of Statistics and Probability [electronic resource]*. Springer Texts in Statistics. Springer Science & Business Media, New York.

Dennis Jr JE and Welsch RE 1978 Techniques for nonlinear least squares and robust regression. *Communications in Statistics-Simulation and Computation* **7** (4), 345–359.

Dolman H, Valentini R and Freibauer A 2008 *The continental-scale greenhouse gas balance of Europe*, vol. 203. Springer Science & Business Media, New York.

Dürre A, Fried R and Liboschik T 2015 Robust estimation of (partial) autocorrelation. *WIREs: Computational Statistics* **7** (3), 205–222.

Edgeworth FY 1887 On observations relating to several quantities. *Hermathena* **6** (13), 279–285.

Emerson JD, Hoaglin DC and Kempthorne PJ 1984 Leverage in least squares additive-plus-multiplicative fits for two-way tables. *Journal of the American Statistical Association* **79** (386), 329.

Etheridge D, Steele L, Francey R and Langenfelds R 1998 Atmospheric methane between 1000 AD and present: Evidence of anthropogenic emissions and climatic variability. *Journal of Geophysical Research* **103** (D13), 15979–15993.

Fox T, Hinkley D and Larntz K 1980 Jackknifing in nonlinear regression. *Technometrics* **22** (1), 29.

Grassia A and De Boer E 1980 Some methods of growth curve fitting. *Math. Scientist* **5**, 91–103.

Habshah M, Norazan M and Rahmatullah Imon A 2009 The performance of diagnostic-robust generalized potentials for the identification of multiple high leverage points in linear regression. *Journal of Applied Statistics* **36** (5), 507–520.

Hadi A 1992 A new measure of overall potential influence in linear-regression. *Computational Statistics and Data Analysis* **14** (1), 1–27.

Hoaglin DC and Welsch RE 1978 The hat matrix in regression and ANOVA. *The American Statistician* **32** (1), 17.

Hoaglin DC, Mosteller F and Tukey JW 1983 *Understanding robust and exploratory data analysis,* vol. 3. John Wiley and Sons Inc., New York.

Huber PJ 1964 Robust estimation of a location parameter. *Annals of Mathematical Statistics* **35** (1), 73 –101.

Huber PJ 1972 The 1972 Wald Lecture. robust statistics: A review. *Annals of Mathematical Statistics* **43** (4), 1041–1067.

Huber PJ 1973 Robust regression: Asymptotics, conjectures and Monte Carlo. *Annals of Statistics* **1** (5), 799–821.

Huber PJ 1981 *Robust statistics.* Wiley Series in Probability and Statistics. John Wiley and Sons.

Huber PJ 1984 Finite sample breakdown of m- and p-estimators. *Annals of Statistics* **12** (1), 119–126.

Imon A 2002 Identifying multiple high leverage points in linear regression. *Journal of Statistical Studies* **3**, 207–218.

Imon A 2005 A stepwise procedure for the identification of multiple outliers and high leverage points in linear regression. *Pakistan Journal of Statistics* **21**, 71–86.

Jennrich RI 1969 Asymptotic properties of non-linear least squares estimators. *Annals of Mathematical Statistics* **71**, 633–643.

Kennedy W and Gentle J 1980 *Statistical Computing.* Statistics: A Series of Textbooks and Monographs. Taylor & Francis.

Koenker R and Bassett G 1978 Regression quantiles. *Econometrica* **46** (1), 33–50.

Koenker R and Park BJ 1996 An interior point algorithm for nonlinear quantile regression. *Journal of Econometrics* **71** (1), 265–283.

Koenker RW and D'Orey V 1987 Algorithm as 229: Computing regression quantiles. *Journal of the Royal Statistical Society. Series C (Applied Statistics)* **36** (3), 383–393.

Lim C, Sen P and Peddada S 2010 Statistical inference in nonlinear regression under heteroscedasticity. *Sankhya B* **72** (2), 202.

Lim C, Sen PK and Peddada SD 2012 Accounting for uncertainty in heteroscedasticity in nonlinear regression. *Journal of Statistical Planning and Inference* **142** (5), 1047–1062.

Maronna R and Yohai V 1981 Asymptotic behavior of general M-estimates for regression and scale with random carriers. *Zeitschrift für Wahrscheinlichkeitstheorie und Verwandte Gebiete* **58** (1), 7–20.

Maronna RA, Martin RD and Yohai VJ 2006 *Robust statistics theory and methods.* Wiley Series in Probability and Statistics. John Wiley and Sons, Chichester.

Martin RD 1980 Robust estimation of autoregressive models. *Directions in Time Series* **1**, 228–262.

Midi H and Jafaar A 2004 The residual plot for a non-linear regression model with the presence of outliers and heteroscedastic errors. *Journal of Technologi* **C** (41), 11–26.

Motulsky H and Brown R 2006 Detecting outliers when fitting data with nonlinear regression – a new method based on robust nonlinear regression and the false discovery rate. *BMC Bioinformatics* **7** (1), 123.

Nocedal J and Wright SJ 2006 *Numerical Optimization.* Springer.

Osborne MR and Watson GA 1971 On an algorithm for discrete nonlinear L1 approximation. *Computer Journal* **14** (2), 184.

Raupach MR, Barrett DJ, Briggs PR *et al.* 2005 Simplicity, complexity and scale in terrestrial biosphere modelling. *Predictions in Ungauged Basins (PUB) workshop, Perth, Australia, 2–5 February 2004*, pp. 239–274 . Food and Agriculture Organization of the United Nations.

Riani M, Cerioli A and Torti F 2014 On consistency factors and efficiency of robust S-estimators. *TEST* **23** (2), 356–387.

Riazoshams H and Midi H 2009 A nonlinear regression model for chickens growth data. *European Journal of Scientific Research.*

Riazoshams H and Midi H 2014 Robust nonlinear regression: case study for modeling the greenhouse gases, methane and carbon dioxide concentration in atmosphere. *Malaysian Journal of Mathematical Sciences* **8** (S), 173–184.

Riazoshams H and Midi HB 2016 The performance of a robust multistage estimator in nonlinear regression with heteroscedastic errors. *Communications in Statistics – Simulation and Computation* **45** (9), 3394–3415.

Riazoshams H and Miri H 2005 Investigating growth models using nonlinear regression models. Technical report, Islamic Azad University, Abade branch, Fars province, Iran.

Riazoshams H, Habshah M and Adam MB 2011 On the outlier detection in nonlinear regression. *World Academy of Science, Engineering and Technology* **60**, 264–270.

Riazoshams H, Midi HB and Sharipov OS 2010 The performance of robust two-stage estimator in nonlinear regression with autocorrelated error. *Communications in Statistics – Simulation and Computation* **39** (6), 1251–1268.

Rousseeuw P and Yohai V 1984 Robust regression by means of S-estimators. In *Robust and Nonlinear Time Series Analysis* (ed. Franke J, Härdle W and Martin D) vol. 26 of *Lecture Notes in Statistics.* Springer, pp. 256–272.

Rousseeuw PJ 1984 Least median of squares regression. *Journal of the American Statistical Association* **79** (388), 871–880.

Rousseeuw PJ 1985 Multivariate estimation with high breakdown point. *Mathematical Statistics and Applications* **8**, 283–297.

Sakata S and White H 2001 S-estimation of nonlinear regression models with dependent and heterogeneous observations. *Journal of Econometrics* **103** (1), 5–72.

Seber GAF and Wild CJ 2003 *Nonlinear Regression Analysis and its Applications.* Wiley Series in Probability and Mathematical Statistics. John Wiley and Sons Inc., New York.

Serfling RJ 2002 *Approximation theorems of mathematical statistics.* Wiley Series in Probability and Mathematical Statistics. John Wiley and Sons Inc., New York.

Shevlyakov G, Lyubomishchenko N and Smirnov P 2013 *Some remarks on robust estimation of power spectra.* BSU Publishing Center, Minsk.

Silvestre A, Petim-Batista F and Colaco J 2006 The accuracy of seven mathematical functions in modeling dairy cattle lactation curves based on test-day records from varying sample schemes. *Journal of Dairy Science* **89** (5), 1813–1821.

Sinha SK, Field CA and Smith B 2003 Robust estimation of nonlinear regression with autoregressive errors. *Statistics and Probability Letters* **63** (1), 49.

Smirnov P and Shevlyakov G 2010 On approximation of the Qn-estimate of scale by fast M-estimates. *Book of Abstracts of the International Conference on Robust Statistics*, pp. 94–95.

Srikantan KS 1961 Testing for the single outlier in a regression model. *Sankhyā: The Indian Journal of Statistics, Series A* **23** (3), 251.

St. Laurent RT and Cook RD 1992 Leverage and superleverage in nonlinear regression. *Journal of the American Statistical Association* **87**, 985–990.

St. Laurent RT and Cook RD 1993 Leverage, local influence and curvature in nonlinear regression. *Biometrika* **80**, 99–106.

Stromberg AJ 1993 Computation of high breakdown nonlinear regression parameters. *Journal of the American Statistical Association* **88** (421), 237–244.

Stromberg AJ 1995 Consistency of the least median of squares estimator in nonlinear regression. *Communications in Statistics – Theory and Methods* **24** (8), 1971–1984.

Stromberg AJ and Ruppert D 1992 Breakdown in nonlinear regression. *Journal of the American Statistical Association* **87** (420), 991–997.

Stromberg AJ, Hössjer O and Hawkins DM 2000 The least trimmed differences regression estimator and alternatives. *Journal of the American Statistical Association* **95** (451), 853–864.

Tabatabai M and Argyros I 1993 Robust estimation and testing for general nonlinear regression models. *Applied Mathematics and Computation* **57** (1), 85–101.

UNEP 1989 *Environmental data report prepared for UNEP by the GEMS Monitoring and Assessment Research Centre*. London, UK, in co-operation with the World Resources Institute, Washington, DC.

United States Environmental Protection Agency 1978 *A Compendium of Lake and Reservoir Data Collected by the National Eutrophication Survey in Eastern North-central, and Southeastern United States*. Working paper (National Eutrophication Survey (US)). Corvallis Environmental Research Laboratory.

Vankeerberghen P, Smeyers-Verbeke J, Leardi R, Karr CL and Massart DL 1995 Robust regression and outlier detection for non-linear models using genetic algorithms. *Chemometrics and Intelligent Laboratory Systems* **28**, 73–87.

Wagner HM 1959 Linear programming techniques for regression analysis. *Journal of the American Statistical Association* **54** (285), 206–212.

Wei BC, Hu YQ and Fung WK 1998 Generalized leverage and its applications. *Scandinavian Journal of Statistics* **25** (1), 25–37.

Wei WW 2006 *Time Series Analysis – Univariate and Multivariate Methods, 2nd edition*. Pearson Addison Wesley, Boston.

White G and Brisbin Jr I 1980 Estimation and comparison of parameters in stochastic growth models for barn owls. *Growth* **44** (2), 97–111.

Yohai VJ 1985 High breakdown-point and high efficiency robust estimates for regression. Technical report, Department of Statistics, University of Washington, Seattle.

Yohai VJ 1987 High breakdown-point and high efficiency robust estimates for regression. *Annals of Statistics* **15** (2), 642–656.

Yohai VJ and Zamar RH 1988 High breakdown-point estimates of regression by means of the minimization of an efficient scale. *Journal of the American Statistical Association* **83** (402), 406–413.

Index

Robust Nonlinear Regression: with Applications using R, First Edition.
Hossein Riazoshams, Habshah Midi, and Gebrenegus Ghilagaber.
© 2019 John Wiley & Sons Ltd. Published 2019 by John Wiley & Sons Ltd.
Companion website: www.wiley.com/go/riazoshams/robustnonlinearregression